THE SCIENCES OF THE ARTIFICIAL

KARL TAYLOR COMPTON *1887–1954*

The Karl Taylor Compton Lectures honor the memory of the ninth president of the Massachusetts Institute of Technology by bringing to the M.I.T. community some of the great minds of our time who contribute to the integration of scientific, cultural, and philosophical concerns—a synthesis richly achieved by Karl Taylor Compton and shared with colleagues and students during his long leadership of the Institute.

KARL TAYLOR COMPTON LECTURERS

1957 Niels Bohr

1959 Otto Struve

1960 André Lwoff

1962 Isidor I. Rabi

1968 Herbert A. Simon

THE M.I.T. PRESS
Massachusetts Institute of Technology
Cambridge, Massachusetts, and London, England

THE SCIENCES OF THE ARTIFICIAL

by Herbert A. Simon

Set in Monotype Times Roman.
Printed and bound in the United States of America
by The Riverside Press, Inc.

Library of Congress catalog card number: 69-11312

TO ALLEN NEWELL
IN FRIENDSHIP

PREFACE

The invitation to deliver the Karl Taylor Compton lectures at the Massachusetts Institute of Technology in the spring of 1968 provided me with a welcome opportunity to make explicit and to develop at some length a thesis that has been central to much of my research, at first in organization theory, later in management science, and most recently in psychology.

The thesis is that certain phenomena are "artificial" in a very specific sense: They are as they are only because of a system's being molded, by goals or purposes, to the environment in which it lives. If natural phenomena have an air of "necessity" about them in their subservience to natural law, artificial phenomena have an air of "contingency" in their malleability by environment.

The contingency of artificial phenomena has always created doubts as to whether they fall properly within the compass of science. Sometimes these doubts are directed at the teleological character of artificial systems and the consequent difficulty of disentangling prescription from description. This seems to me not to be the real difficulty. The genuine problem is to show how empirical propositions can be made at all about systems that, given different circumstances, might be quite other than they are.

Almost as soon as I began research on administrative organizations, some thirty years ago, I encountered the problem of artificiality in almost its pure form:

> . . . administration is not unlike play-acting. The task of the good actor is to know and play his role, although different roles may differ greatly in content. The effectiveness of the performance will depend on the effectiveness of the play and the effectiveness with which it is played. The effectiveness of the administrative process will vary with the effectiveness of the organization and the effectiveness with which its members play their parts. [*Administrative Behavior*, p. 252].

How, then, could one construct a theory of administration that would contain more than the normative rules of good acting? In particular, how could one construct an empirical theory? My writing on administration, particularly in *Administrative Behavior* and Part IV of *Models of Man*, has sought to answer those questions by showing that the empirical content of the phenomena, the necessities that rise above the contingencies, stems from the inabilities of the behavioral system to adapt perfectly to its environment—from the limits of rationality, as I have called them.

As research took me into other areas, it became evident that the problem of artificiality was not peculiar to administration and organizations but that it infected a far wider range of subjects. Economics, since it postulated rationality in economic man, made him the supremely skillful actor, whose behavior could reveal something of the requirements the environment placed on him but nothing about his own cognitive make-up. But the difficulty must then extend beyond economics into all those parts of psychology that are concerned with rational behavior—thinking, problem solving, learning.

Finally, I thought I began to see in the problem of artificiality an explanation of the difficulty that has been experienced in filling engineering and other professions with empirical and theoretical substance distinct from the substance of their supporting sciences. Engineering, medicine, business, architecture, and painting are concerned not with the necessary but with the contingent—not with how things are but with how they might be—in short, with design, The possibility of creating a science or sciences of design is exactly as great as the possibility of creating any science of the artificial. The two possibilities stand or fall together.

These essays, then, attempt to explain how a science of the artificial is possible and to illustrate its nature. I have not drawn my illustrations from administration, organization theory, or economics, since I have discussed these subjects at length elsewhere. Instead, I have taken as my main examples—in the second and third lectures, respectively—the fields of the psychology of cognition and engineering design. Since Karl Compton was a distinguished engineering educator as well as a distinguised scientist, I thought it not inappropriate to apply my conclusions about design to the question of reconstructing the engineering curriculum.

The reader will discover, in the course of discussion, that artificiality is interesting principally when it concerns complex systems that live in complex environments. The topics of artificiality and complexity are inextricably interwoven. For this reason, I have included in this volume an earlier essay, "The Architecture of Complexity," which develops at length some ideas about complexity that I could touch on only briefly in these lectures. The essay appeared originally in the December 1962 *Proceedings of the American Philosophical Society*.

I have tried to acknowledge some specific debts to others in footnotes at appropriate points in the text. I owe a much more general debt to Allen Newell, whose partner I have been in a very large part of my work for more than a decade, and to whom I have dedicated this volume. If there are parts of my thesis with which he disagrees, they are probably wrong; but he cannot evade a major share of responsibility for the rest.

Lee W. Gregg will recognize many ideas, particularly in the second chapter, as having had their origins in work we have done together; and other colleagues, as well as numerous present and former graduate students, have left their fingerprints on various pages of the text. Among the latter, I want to mention specifically L. Stephen Coles, Edward A. Feigenbaum, John Grason, Robert K. Lindsay, Ross Quillian, Laurent Siklossy, Donald S. Williams, and Thomas G. Williams, whose work is particularly relevant to the topics discussed here.

Previous versions of Chapter Four incorporated valuable suggestions and data contributed by George W. Corner, Richard H. Meier, John. R. Platt, Andrew Schoene, Warren Weaver, and William Wise.

A large part of the psychological research reported in the second chapter was supported by the Public Health Service Research Grant MH-07722 from the National Institutes of Mental Health and some of the research on design reported in the third chapter by the Advanced Research Projects Agency of the Office of the Secretary of Defense (SD-146). These grants, as well as earlier support from the Carnegie Corporation and the Ford Foundation, have enabled us at Carnegie-Mellon to pursue for over a decade a many-pronged exploration aimed at deepening our understanding of artificial phenomena.

Finally, I am grateful to the Massachusetts Institute of Technology for the opportunity to prepare and present these lectures, and for the occasion to become better acquainted with the research in the sciences of the artificial that is now going forward on the M.I.T. campus.

Herbert A. Simon

Pittsburgh, Pennsylvania
April 2, 1968

CONTENTS

1 UNDERSTANDING THE NATURAL AND THE ARTIFICIAL WORLDS

About three centuries after Newton, we are thoroughly familiar with the concept of natural science—most unequivocally with physical and biological science. A natural science is a body of knowledge about some class of things—objects or phenomena—in the world: about the characteristics and properties that they have; about how they behave and interact with each other.

The central task of a natural science is to make the wonderful commonplace: to show that complexity, correctly viewed, is only a mask for simplicity; to find pattern hidden in apparent chaos. The early Dutch physicist Simon Stevin, showed by an elegant drawing (Figure 1) that the law of the inclined plane follows in "self-evident fashion" from the impossibility of perpetual motion,

Figure 1. The Law of the Inclined Plane
The vignette devised by Simon Stevin, to illustrate his derivation of the law.

for experience and reason tell us that the chain of balls in the figure would rotate neither to right nor to left but would remain at rest. (Since rotation changes nothing in the figure, if the chain moved at all, it would move perpetually.) Since the pendant part of the chain hangs symmetrically, we can snip it off without disturbing the equilibrium. But now the balls on the long side of the plane balance those on the shorter, steeper side; and their relative numbers are in inverse ratio to the sines of the angles at which the planes are inclined.

Stevin was so pleased with his construction that he incorporated it into a vignette, inscribing above it

WONDER, EN IS GHEEN WONDER

that is to say: "Wonderful, but not incomprehensible."

This is the task of natural science: to show that the wonderful is not incomprehensible, to show how it can be comprehended—but not to destroy wonder. For when we have explained the wonderful, unmasked the hidden pattern, a new wonder arises at how complexity was woven out of simplicity. The aesthetics of natural science and mathematics is at one with the aesthetics of music and painting—both inhere in the discovery of a partially concealed pattern.

The world we live in today is much more a man-made, or artificial, world than it is a natural world. Almost every element in our environment shows evidence of man's artifice. The temperature in which we spend most of our hours is kept artificially at 70 degrees; the humidity is added to or taken from the air we breathe; and the impurities we inhale are largely produced (and filtered) by man.

Moreover, for most of us—the white-collared ones—the significant part of the environment consists mostly of strings of artifacts called "symbols" that we receive through eyes and ears in the form of written and spoken language and that we pour out into the environment—as I am now doing—by mouth or hand. The laws that govern these strings of symbols, the laws that govern the occasions on which we emit and receive them, the determinants of their content are all consequences of our collective artifice.

One may object that I exaggerate the artificiality of our world. Man must obey the law of gravity as surely as does a stone; and he is a living organism who depends, for his food and in many other ways, on the world of biological phenomena. I shall plead guilty to overstatement, while protesting that the exaggeration is slight. To say that an astronaut, or even an airplane pilot, is obeying the law of gravity, hence is a perfectly natural phenomenon, is true; but its truth calls for some sophistication in what we mean by "obeying" a natural law. Aristotle did not think it natural for heavy things to rise or light ones to fall (*Physics*, Book IV); but presumably we have a deeper understanding of "natural" than he did.

So, too, we must be careful about equating "biological" with "natural." A forest may be a phenomenon of nature; a farm certainly is not. The very species upon which man depends for his food—his corn and his cattle—are artifacts of his ingenuity. A plowed field is no more part of nature than an asphalted street—and no less.

These examples set the terms of our problem, for those things we call artifacts are not apart from nature. They have no dispensation to ignore or violate natural law. At the same time, they are adapted to man's goals and purposes. They are what they are in order to satisfy his desire to fly or to eat well. As man's aims change, so too do his artifacts—and vice versa, as well.

If science is to encompass these objects and phenomena in which human purpose as well as natural law are embodied, it must have means for relating these two disparate components. The character of these means and their implications for certain areas of knowledge—psychology and engineering in particular—are the central concern of the first three chapters.

The Artificial

Natural science is knowledge about natural objects and phenomena. We ask whether there cannot also be "artificial" science—knowledge about artificial objects and phenomena. Unfortunately, the term "artificial" has a pejorative air about it that we must dispel before we can proceed.

My dictionary defines "artificial" as: "Produced by art rather than by nature; not genuine or natural; affected; not pertaining to the essence of the matter." It proposes, as synonyms: affected, factitious, manufactured, pretended, sham, simulated, spurious, trumped up, unnatural. As antonyms, it lists: actual, genuine, honest, natural, real, truthful, unaffected. Our language seems to reflect man's deep distrust of his own products. I shall not try to assess the validity of that evaluation or explore its possible psychological roots. But you will have to understand me as using "artificial" in as neutral a sense as possible: as meaning man-made, as opposed to natural.[1]

In some contexts, we make a distinction between "artificial" and "synthetic." For example, a gem made of glass colored to resemble sapphire would be called

[1] I shall disclaim responsibility for this particular choice of terms. The phrase "artificial intelligence," which led me to it, was coined, I think, right on the River Charles, at M.I.T. Our own research group at RAND and Carnegie-Mellon University have preferred phrases like "complex information processing" and "simulation of cognitive processes." But then we run into new terminological difficulties, for the dictionary also says that "to simulate" means "to assume or have the mere appearance or form of, without the reality; imitate; counterfeit; pretend." At any rate, "artificial intelligence" seems to be here to stay, and it may prove easier to cleanse the phrase than to dispense with it. In time it will become sufficiently idiomatic that it will no longer be the target of cheap rhetoric.

artificial, while a man-made gem that was chemically indistinguishable from sapphire would be called synthetic. A similar distinction is often made between "artificial" and "synthetic" rubber. Thus some artificial things are imitations of things in nature; and the imitation may use either the same basic materials as those in the natural object or quite different materials.

As soon as we introduce "synthesis" as well as "artifice" we enter the realm of engineering. For "synthetic" is often used in the broader sense of "designed" or "composed." We speak of engineering as concerned with "synthesis," while science is concerned with "analysis." Synthetic or artificial objects—and more specifically, prospective artificial objects having desired properties—are the central objective of engineering activity and skill. The engineer is concerned with how things *ought* to be—ought to be, that is, in order to *attain goals*, and to *function*. Hence, a science of the artificial will be closely akin to a science of engineering—but very different, as we shall see in my third chapter, from what goes currently by the name of "engineering science."

With goals and "oughts," we also introduce into the picture the dichotomy between normative and descriptive. Natural science has found a way to exclude the normative and to concern itself solely with how things are. Can or should we maintain this exclusion when we move from natural to artificial phenomena, from analysis to synthesis?[2]

We have now identified four indicia that distinguish the artificial from the natural; hence we can set the boundaries for sciences of the artificial:

1. Artificial things are synthesized (though not always or usually with full forethought) by man.
2. Artificial things may imitate appearances in natural things while lacking, in one or many respects, the reality of the latter.

[2] This issue will also be discussed at length in my third chapter. In order not to keep readers in suspense, I may say that I hold to the pristine positivist position of the irreducibility of "ought" to "is," as in Chapter III of my *Administrative Behavior* (New York: Macmillan, 1947). This position is entirely consistent with treating natural or artificial goal-seeking systems as phenomena, without commitment to their goals. *Ibid.*, Appendix. See also the well-known paper by A. Rosenbluth, N. Wiener, and J. Bigelow, "Behavior, Purpose, and Teleology," *Philosophy of Science*, *10*: 18–24 (1943).

3. Artificial things can be characterized in terms of functions, goals, adaptation.
4. Artificial things are often discussed, particularly when they are being designed, in terms of imperatives as well as descriptives.

The Environment as Mold

Let us look a little more closely at the functional or purposeful aspect of artificial things. Fulfillment of purpose or adaptation to a goal involves a relation among three terms: the purpose or goal, the character of the artifact, and the environment in which the artifact performs. When we think of a clock, for example, in terms of purpose, we may use the child's definition: "a clock is to tell time." When we focus our attention on the clock itself, we may describe it in terms of arrangements of gears, and of the application of the forces of springs or gravity operating on a weight or pendulum.

But we may also consider clocks in relation to the environment in which they are to be used. Sundials perform as clocks *in sunny climates*—they are more useful in Phoenix than in Boston, and of no use at all during the Arctic winter. Devising a clock that would tell time, on a rolling and pitching ship, with sufficient accuracy to determine longitude was one of the great adventures of eighteenth-century science and technology. To perform in this difficult environment, the clock had to be endowed with many delicate properties, some of them largely or totally irrelevant to the performance of a landlubber's clock.

Natural science impinges on an artifact through two of the three terms of the relation that characterizes it: the structure of the artifact itself and the environment in which it performs. Whether a clock will in fact tell time depends on its internal construction and where it is placed. Whether a knife will cut depends on the material of its blade and the hardness of the substance to which it is applied.

The Artifact as "Interface"

We can view the matter quite symmetrically. An artifact

can be thought of as a meeting point—an "interface" in today's terms—between an "inner" environment, the substance and organization of the artifact itself, and an "outer" environment, the surroundings in which it operates. If the inner environment is appropriate to the outer environment, or vice versa, the artifact will serve its intended purpose. Thus, if the clock is immune to buffeting, it will serve as a ship's chronometer. (And conversely, if it isn't, we may salvage it by mounting it on the mantel at home.)

Notice that this way of viewing artifacts applies equally well to many things that are not man-made—to all things, in fact, that can be regarded as "adapted" to some situation; and in particular, it applies to the living systems that have evolved through the forces of organic evolution. A theory of the airplane draws on natural science for an explanation of its inner environment (the power plant, for example), its outer environment (the character of the atmosphere at different altitudes), and the relation between its inner and outer environments (the movement of an airfoil through a gas). But a theory of the bird can be divided up in exactly the same way.[3]

Given an airplane, or *given* a bird, we can analyze them by the methods of natural science without any particular attention to purpose or adaptation, without reference to the interface between what I have called the inner and outer environments. After all, their behavior is governed by natural law just as fully as the behavior of anything else (or, at least, we all believe this about the airplane, and most of us believe it about the bird).

Functional Explanation

On the other hand, if the division between inner and outer environment is not necessary to the analysis of an airplane or a bird, it turns out at least to be highly convenient.

[3] A generalization of the argument made here for the separability of "outer" from "inner" environment shows that we should expect to find this separability, to a greater or lesser degree, in *all* large and complex systems, whether they are artificial or natural. In its generalized form, it is an argument that all nature will be organized in "levels." My essay "The Architecture of Complexity," included in this volume, develops the more general argument in some detail.

There are several reasons for this, which will become evident from examples.

Many animals in the Arctic have white fur. We usually explain this by saying that white is the best color for the Arctic environment, for white creatures escape detection more easily than do others. This is not, of course, a natural science explanation; it is an explanation by reference to purpose or function. It simply says that these are the kinds of creatures that will "work," that is, survive, in this kind of environment. To turn the statement into an explanation, we must add to it a notion of natural selection, or some equivalent mechanism.

An important fact about this kind of explanation is that it demands an understanding mainly of the outer environment. Looking at our snowy surroundings, we can predict the predominant color of the creatures we are likely to encounter; we need know little about the biology of the creatures themselves, beyond the facts that they are often mutually hostile, use visual clues to guide their behavior, and are adaptive (through selection or some other mechanism).

Analogous to the role played by natural selection in evolutionary biology is the role played by rationality in the sciences of human behavior. If we know of a business organization only that it is a profit-maximizing system, we can often predict how its behavior will change if we change its environment—how it will alter its prices if a sales tax is levied on its products. We can make this prediction—and economists do make it repeatedly—without any detailed assumptions about the adaptive mechanism, the decision-making apparatus that constitutes the inner environment of the business firm.

Thus the first advantage of dividing outer from inner environment in studying an adaptive or artificial system is that we can often predict behavior from knowledge of the system's goals and its outer environment, with only minimal assumptions about the inner environment. An instant corollary is that we often find quite different inner environments accomplishing identical or similar goals in identical or similar outer environments—airplanes and birds, dolphins and tunafish, weight-driven clocks and spring-driven clocks, electrical relays and transistors.

There is often a corresponding advantage in the division from the standpoint of the inner environment. In very many cases, whether a particular system will achieve a particular goal or adaptation depends on only a few characteristics of the outer environment, and not at all on the detail of that environment. Biologists are familiar with this property of adaptive systems under the label of homeostasis. It is an important property of most good designs, whether biological or artifactual. In one way or another, the designer insulates the inner system from the environment, so that an invariant relation is maintained between inner system and goal, independent of variations over a wide range in most parameters that characterize the outer environment. The ship's chronometer reacts to the pitching of the ship only in the negative sense of maintaining an invariant relation of the hands on its dial to the real time, independently of the ship's motions.

Quasi independence from the outer environment may be maintained by various forms of passive insulation, by reactive negative feedback (the most frequently discussed form of insulation), by predictive adaptation, or by various combinations of these.

Functional Description and Synthesis

In the best of all possible worlds—at least, for a designer—we might even hope to combine the two sets of advantages we have described that derive from factoring an adaptive system into goals, outer environment, and inner environment. We might hope to be able to characterize the main properties of the system and its behavior without elaborating the detail of *either* the outer or inner environments. We might look toward a science of the artificial that would depend on the relative simplicity of the interface as its primary source of abstraction and generality.

Consider the design of a physical device to serve as a counter. If we want the device to be able to count up to one thousand, say, it must be capable of assuming any one of at least a thousand states, of maintaining itself in any given state, and of shifting from any state to the "next" state. There are dozens of different inner environments that might be used (and have been used) for such a device. A wheel, notched at each twenty minutes of arc, and with a ratchet device to turn and hold it would do the trick. So would a string of ten electrical switches, properly connected to represent binary numbers. Instead of switches, today we are likely to use transistors, or other solid-state devices.[4]

Our counter would be activated by some kind of pulse, mechanical or electrical, as appropriate, from the outer environment. But by building an appropriate transducer between the two environments, the physical character of the interior pulse could again be made independent of the physical character of the exterior pulse—the counter could be made to count anything.

Description of an artifice in terms of its organization and functioning—its interface between inner and outer environments—is a major objective of invention and design activity. Engineers will find familiar the language of the following claim quoted from a 1919 patent on an improved motor controller:

> What I claim as new and desire to secure by Letters Patent is:
> 1. In a motor controller, in combination, reversing means, normally effective field-weakening means and means associated with said reversing means for rendering said field-weakening means ineffective during motor starting and thereafter effective to different degrees determinable by the setting of said reversing means. . . . [5]

Apart from the fact that we know the invention relates to control of an electric motor, there is almost no reference here to specific, concrete objects or phenomena. There is reference rather to "reversing means" and "field-weakening means," whose further purpose is made clear in a paragraph preceding the patent claims:

> The advantages of the special type of motor illustrated and the control thereof will be readily understood by those skilled in the art. Among such advantages may be mentioned the

[4] The theory of functional equivalence of computing machines has had considerable development in recent years. See Marvin L. Minsky, *Computation: Finite and Infinite Machines* (Englewood Cliffs, N. J.: Prentice-Hall, 1967), Chapters 1–4.

[5] U.S. Patent 1,307,836, granted to Arthur Simon, June 24, 1919.

provision of a high starting torque and the provision for quick reversals of the motor.[6]

Now let us suppose that the motor in question is incorporated in a planing machine (see Figure 2). The inventor describes its behavior thus:

> Referring now to [Figure 2], the controller is illustrated in outline connection with a planer (100) operated by a motor M, the controller being adapted to govern the motor M and to be automatically operated by the reciprocating bed (101) of the planer. The master shaft of the controller is provided with a lever (102) connected by a link (103) to a lever (104) mounted upon the planer frame and projecting into the path of lugs (105) and (106) on the planer bed. As will be understood, the arrangement is such that reverse movements of the planer bed will, through the connections described, throw the master shaft of the controller back and forth between its extreme positions and in consequence effect selective operation of the reversing switches (1) and (2) and automatic operation of the other switches in the manner above set forth.[7]

In this manner, the properties with which the inner environment has been endowed are placed at the service of the goals in the context of the outer environment. The motor will reverse periodically under the control of the position of the planer bed. The "shape" of its behavior—the time path, say, of a variable associated with the motor—will be a function of the "shape" of the external environment—the distance, in this case, between the lugs on the planer bed.

The device we have just described illustrates in microcosm the nature of artifacts. Central to their description are the goals that link the inner to the outer system. The inner system is an organization of natural phenomena capable of attaining the goals in some range of environments; but ordinarily there will be many functionally equivalent natural systems capable of doing this.

The outer environment determines the conditions for goal attainment. If the inner system is properly designed, it will be adapted to the outer environment, so that its behavior will be determined in large part by the behavior of the latter, exactly as in the case of "economic man." To predict how it will behave, we need only ask "How

[6] *Ibid.*
[7] *Ibid.*

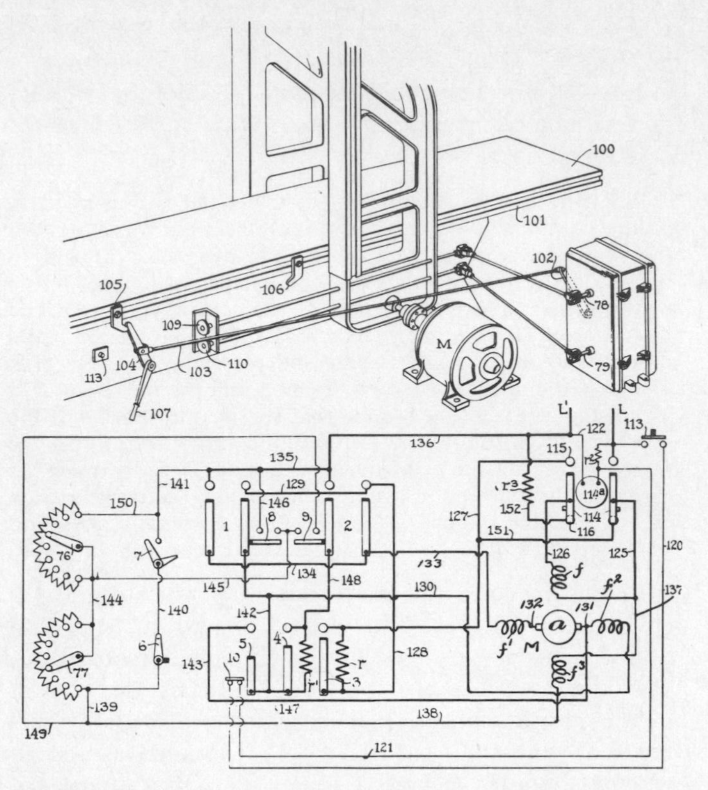

Figure 2. Illustrations from a Patent for a Motor Controller

would a rationally designed system behave under these circumstances?" The behavior takes on the shape of the task environment.[8]

Limits of Adaptation

But matters must be just a little more complicated than this account suggests. "If wishes were horses, all beggars would ride." And if we could always specify a Protean inner system that would take on exactly the shape of the task environment, designing would be synonymous with

[8] On the crucial role of adaptation or rationality—and their limits—for economics and organization theory, see the introduction to Part IV, "Rationality and Administrative Decision Making," of my *Models of Man* (New York: Wiley, 1957), and pp. 38–41, 80–81, and 240–244 of *Administrative Behavior*, *op. cit.*

wishing. "Means for scratching diamonds" defines a design objective, an objective that *might* be attained with the use of many different substances. But the design has not been achieved until we have discovered at least one realizable inner system obeying the ordinary natural laws—one material, in this case, hard enough to scratch diamonds.

Often, we shall have to be satisfied with meeting the design objectives only approximately. Then the properties of the inner system will "show through." That is, the behavior of the system will only partly respond to the task environment; partly, it will respond to the limiting properties of the inner system.

Thus, the motor controls described earlier are aimed at providing for "quick" reversal of the motor. But the motor must obey electromagnetic and mechanical laws, and we could easily confront the system with a task where the environment called for quicker reversal than the motor was capable of. In a benign environment we would learn from the motor only what it had been called upon to do; in a taxing environment we would learn something about its internal structure—specifically, about those aspects of the internal structure that were chiefly instrumental in limiting performance.[9]

A bridge, under its usual conditions of service, behaves simply as a relatively smooth level surface on which vehicles can move. Only when it has been overloaded do we learn the physical properties of the materials from which it is built.

Understanding by Simulating

Artificiality connotes perceptual similarity but essential difference, resemblance from without rather than within.

[9] Compare the corresponding proposition on the design of administrative organizations: "Rationality, then, does not determine behavior. Within the area of rationality behavior is perfectly flexible and adaptable to abilities, goals, and knowledge. Instead, behavior is determined by the irrational and nonrational elements that bound the area of rationality . . . administrative theory must be concerned with the limits of rationality, and the manner in which organization affects these limits for the person making a decision." *Administrative Behavior*, *op. cit.*, p. 241.

In the terms of the previous section, we may say that the artificial object imitates the real by turning the same face to the outer system, by adapting, relative to the same goals, to comparable ranges of external tasks. Imitation is possible because distinct physical systems can be organized to exhibit nearly identical behavior. The damped spring and the damped circuit obey the same second-order linear differential equation; hence we may use either one to imitate the other.

Techniques of Simulation

Because of its abstract character and its symbol-manipulating generality, the digital computer has greatly extended the range of systems whose behavior can be imitated. Generally, we now call the imitation "simulation," and we try to understand the imitated system by testing the simulation in a variety of simulated, or imitated, environments.

Simulation, as a technique for achieving understanding and predicting the behavior of systems, predates, of course, the digital computer. The model basin and the wind tunnel are valued means for studying the behavior of large systems by modeling them in the small; and it is quite certain that Ohm's law was suggested to its discoverer by its analogy with simple hydraulic phenomena.

Simulation may even take the form of a thought experiment, never actually implemented dynamically. One of my vivid memories of the Great Depression is of a large multicolored chart in my father's study that represented a hydraulic model of an economic system (with different fluids for money and goods). The chart was devised, I believe, by a technocratically inclined engineer named Dahlberg. The model never got beyond the pen-and-paint stage at that time, but it could be used to trace through the imputed consequences of particular economic measures or events—provided the theory was right!

As my formal education in economics progressed, I acquired a disdain for that naïve simulation, only to discover after the Second World War that a distinguished economist, Professor Abba Lerner, had actually built the Moniac, a hydraulic model that simulated a Keynesian economy. Of course, Professor Lerner's simulation incorporated a more nearly correct theory than the earlier one, and was actually constructed and operated—two points in its favor. However, the Moniac, while useful as a teaching tool, told us nothing that could not be extracted readily from simple mathematical versions of Keynesian theory, and was soon priced out of the market by the growing number of computer simulations of the economy.

Simulation as a Source of New Knowledge

This brings me to the crucial question about simulation: *How can a simulation ever tell us anything that we do not already know?* The usual implication of the question is that it can't. As a matter of fact, there is an interesting parallelism, which I shall exploit presently, between two assertions about computers and simulation that one hears frequently:

1. A simulation is no better than the assumptions built into it.
2. A computer can do only what it is programmed to do.

I shall not deny either assertion, for both seem to me to be true. But despite both assertions, simulation can tell us things we do not already know.

There are two related ways in which simulation can provide new knowledge—one of them obvious, the other perhaps a bit subtle. The obvious point is that, even when we have correct premises, it may be very difficult to discover what they imply. All correct reasoning is a grand system of tautologies, but only God can make direct use of that fact. The rest of us must painstakingly and fallibly tease out the consequences of our assumptions.

Thus, we might expect simulation to be a powerful technique for deriving, from our knowledge of the mechanisms governing the behavior of gases, a theory of the weather and a means for weather prediction. Indeed, as many people are aware, attempts have been under way for several years to apply this technique. Greatly oversimplified, the idea is that we already know the correct basic assumptions, the local atmospheric equations; but we need the computer to work out the implications of the interactions of vast numbers of variables starting from complicated initial conditions. This is simply an extrapolation to the scale of modern computers of the idea we use when we solve two simultaneous equations by algebra.

This approach to simulation has numerous applications to engineering design. For it is typical of many kinds of design problems that the inner system consists of components whose fundamental laws of behavior—mechanical, electrical, or chemical—are well known. The difficulty of the design problem often resides in predicting how an assemblage of such components will behave.

Simulation of Poorly Understood Systems

The more interesting and subtle question is whether simulation can be of any help to us when we do not know very much, initially, about the natural laws that govern the behavior of the inner system. Let me show why this question must also be answered in the affirmative.

First, I shall make a preliminary comment that simplifies matters: We are seldom interested in explaining or predicting phenomena in all their particularity; we are usually interested only in a few properties abstracted from the complex reality. Thus, a man-launched satellite is surely an artificial object, but we usually do not think of it as "simulating" the moon or a planet. It simply obeys the same laws of physics, which relate only to its inertial and gravitational mass, abstracted from most of its other properties. It *is* a moon. Similarly, electric energy that enters my house from the atomic generating station at Shippingport does not "simulate" energy generated by means of a coal plant or a windmill. Maxwell's equations hold for both.

The more we are willing to abstract from the detail of a set of phenomena, the easier it becomes to simulate the phenomena. Moreover, we do not have to know, or guess at, all the internal structure of the system, but only that part of it that is crucial to the abstraction.

It is fortunate that this is so, for if it were not, the top-down strategy that built the natural sciences over the past three centuries would have been infeasible. We knew a great deal about the gross physical and chemical behavior of matter before we had a knowledge of molecules, a great deal about molecular chemistry before we had an atomic theory, and a great deal about atoms before we had any theory of elementary particles—if, indeed, we have such a theory today.

This skyhook-skyscraper construction of science from the roof down to the yet unconstructed foundations was possible because the behavior of the system at each level depended on only a very approximate, simplified, abstracted characterization of the system at the level next beneath.[10] This is lucky, else the safety of bridges and airplanes might depend on the correctness of the "Eightfold Way" of looking at elementary particles.

Artificial systems and adaptive systems have properties that make them particularly susceptible to simulation via simplified models. The characterization of such systems in the previous section of this chapter explains why. Resemblance in behavior of systems without identity of the inner systems is particularly feasible if the aspects in which we are interested arise out of the *organization* of the parts, independently of all but a few properties of the individual components. Thus, for many purposes, we may be interested in only such characteristics of a material as its tensile and compressive strength. We may be profoundly unconcerned about its chemical properties, or even whether it is wood or iron.

The motor control patent cited earlier illustrates this abstraction to organizational properties. The invention

[10] This point is developed more fully in "The Architecture of Complexity" in this volume. More than fifty years ago, Bertrand Russell made the same point about the architecture of mathematics. See the "Preface" to *Principia Mathematica*: " . . . the chief reason in favour of any theory on the principles of mathematics must always be inductive, i.e., it must lie in the fact that the theory in question enables us to deduce ordinary mathematics. In mathematics, the greatest degree of self-evidence is usually not to be found quite at the beginning, but at some later point; hence the early deductions, until they reach this point, give reasons rather for believing the premises because true consequences follow from them, than for believing the consequences because they follow from the premises." Contemporary preferences for deductive formalisms frequently blind us to this important fact, which is no less true today than it was in 1910.

consisted of a "combination" of "reversing means," of "field-weakening means," that is to say, of components specified in terms of their functioning in the organized whole. How many ways are there of reversing a motor, or of weakening its field strength? We can simulate the system described in the patent claims in many ways without reproducing even approximately the actual physical device that is depicted. With a small additional step of abstraction, the patent claims could be restated to encompass mechanical as well as electrical devices. I suppose that any undergraduate engineer at Carnegie-Mellon University or M.I.T. could design a mechanical system embodying reversibility and variable starting torque so as to simulate the system of the patent.

The Computer as Artifact

No artifact devised by man is so convenient for this kind of functional description as a digital computer. It is truly Protean, for almost the only ones of its properties that are detectable in its behavior (when it is operating properly!) are the organizational properties. The speed with which it performs its basic operations may allow us to infer a little about its physical components and their natural laws; speed data, for example, would allow us to rule out certain kinds of "slow" components. For the rest, almost no interesting statement that one can make about an operating computer bears any particular relation to the specific nature of the hardware. A computer is an organization of elementary functional components in which, to a high approximation, only the function performed by those components is relevant to the behavior of the whole system.[11]

Computers as Abstract Objects

This highly abstractive quality of computers makes it easy to introduce mathematics into the study of their

[11] On the subject of this and the following paragraphs, see M. L. Minsky, *op. cit.*; then, John von Neumann, "Probabilistic Logics and the Synthesis of Reliable Organisms from Unreliable Components," in C. E. Shannon and J. McCarthy (eds.), *Automata Studies* (Princeton: Princeton University Press, 1956).

theory—and has led some to the erroneous conclusion that, as a computer science emerges, it will necessarily be a mathematical rather than an empirical science. Let me take up these two points in turn: the relevance of mathematics to computers, and the possibility of studying computers empirically.

Some important theorizing, initiated by John von Neumann, has been done on the topic of computer reliability. The question is how to build a reliable system from unreliable parts. Notice that this is not posed as a question of physics or physical engineering. The components engineer is assumed to have done his best, but the parts are still unreliable! We can cope with the unreliability only by our manner of organizing them.

To turn this into a meaningful problem, we have to say a little more about the nature of the unreliable parts. Here we are aided by the knowledge that *any* computer can be assembled out of a small array of simple, basic elements. For instance, we may take as our primitives the so-called Pitts–McCulloch neurons. As their name implies, these components were devised in analogy to the supposed anatomical and functional characteristics of neurons in the brain; but they are highly abstracted. They are formally isomorphic with the simplest kinds of switching circuits—"and," "or," and "not" circuits. We postulate, now, that we are to build a system from such elements and that each elementary part has a specified probability of functioning incorrectly. The problem is to arrange the elements and their interconnections in such a way that the complete system will perform reliably.

The important point, for our present discussion, is that the parts could as well be neurons as relays, as well relays as transistors. The natural laws governings relays are very well known, while the natural laws governing neurons are known most imperfectly. But that does not matter, for all that is relevant for the theory is that the components have the specified level of unreliability and be interconnected in the specified way.

This example shows that the possibility of building a mathematical theory of a system or of simulating that system does not depend on having an adequate microtheory of the natural laws that govern the system components. Such a microtheory might indeed be simply irrelevant.

Computers as Empirical Objects

We turn next to the feasibility of an *empirical* science of computers—as distinct from the solid-state physics or physiology of their componentry. As a matter of empirical fact, almost all of the computers that have been designed have certain common organizational features. They almost all can be decomposed into an active processor (Babbage's "Mill") and a memory (Babbage's "Store") in combination with input and output devices. (Some of the larger systems, somewhat in the manner of colonial algae, are assemblages of smaller systems having some or all of these components. But perhaps I may oversimplify for the moment.) They are all capable of storing symbols (program) that can be interpreted by a program-control component and executed. Almost all have exceedingly limited capacity for simultaneous, parallel activity—they are basically one-thing-at-a-time systems. Symbols generally have to be moved from the larger memory components into the central processor before they can be acted upon. The systems are capable of only simple basic actions: recoding symbols, copying symbols, moving symbols, erasing symbols, and comparing symbols.

Since there are now many such devices in the world, and since the properties that describe them also appear to be shared by the human central nervous system, nothing prevents us from developing a natural history of them. We can study them as we would rabbits or chipmunks and discover how they behave under different patterns of environmental stimulation. Insofar as their behavior reflects largely the broad functional characteristics we have described, and is independent of details of their hardware, we can build a general—but empirical—theory of them.

The research that has been done over the past five years on the design of computer time-sharing systems is a good example of the study of computer behavior as an empirical phenomenon. Only fragments of theory are available to guide the design of a time-sharing system or to predict how a system of a specified design will actually behave in an environment of users who place their several demands upon it. Most actual designs have turned out initially to exhibit serious deficiencies; and most predictions of performance have been startlingly inaccurate.

Under these circumstances, the main route open to the development and improvement of time-sharing systems is to build them and see how they behave. And this is what has been done. They have been built, modified, and improved in successive stages. Perhaps theory could have anticipated these experiments and made them unnecessary. In fact, it didn't; and I don't know anyone intimately acquainted with these exceedingly complex systems who has very specific ideas as to how it might have done so. To understand them, the systems had to be constructed, and their behavior observed.[12]

In a similar vein, computer programs designed to play games or to discover proofs for mathematical theorems spend their lives in exceedingly large and complex task environments. Even when the programs themselves are only moderately large and intricate (compared, say, with the monitor and operating systems of large computers), too little is known about their task environments to permit accurate prediction of how well they will perform, how selectively they will be able to search for problem solutions.

Here again, theoretical analysis must be accompanied by large amounts of experimental work. A growing literature reporting these experiments is beginning to give us precise knowledge about the degree of heuristic power of particular heuristic devices in reducing the size of the problem spaces that must be searched. In theorem proving, for example, there has been a whole series of advances in heuristic power based on and guided by empirical exploration: the use of the Herbrand theorem, the resolution principle, the set-of-support principle, and so on.[13]

[12] The empirical, exploratory flavor of computer research is nicely captured by the account of Maurice V. Wilkes in his 1967 Turing Lecture, "Computers Then and Now," *Journal of the Association for Computing Machinery*, *15*: 1–7 (January 1968).

[13] Note, for example, the empirical data in Lawrence Wos, George A. Robinson, Daniel F. Carson, and Leon Shalla, "The Concept of Demodulation in Theorem Proving," *Journal of the Association for Computing Machinery*, *14*: 698–709 (October 1967), and in several

COMPUTERS AND THOUGHT

As we succeed in broadening and deepening our knowledge—theoretical and empirical—about computers, we shall discover that in large part their behavior is governed by simple general laws, that what appeared as complexity in the computer program was, to a considerable extent, complexity of the environment to which the program was seeking to adapt its behavior.

To the extent that this prospect can be realized, it opens up an exceedingly important role for computer simulation as a tool for achieving a deeper understanding of human behavior. For if it is the organization of components, and not their physical properties, that largely determines behavior, and if computers are organized somewhat in the image of man, then the computer becomes an obvious device for exploring the consequences of alternative organizational assumptions for human behavior. Psychology can move forward without awaiting the solutions by neurology of the problems of component design—however interesting and significant these components turn out to be. I shall devote my next chapter to that most interesting of all artificial systems, the human mind, and to the topic of computer simulation of human thinking.

---

of the earlier papers referenced there. See also the collection of programs in Edward Feigenbaum and Julian Feldman (eds.), *Computers and Thought* (New York: McGraw-Hill, 1963). It is common practice in the field to title papers about heuristic programs, "Experiments with an *XYZ* Program."

2 THE PSYCHOLOGY OF THINKING:
Embedding Artifice in Nature

We watch an ant make his laborious way across a wind- and wave-molded beach. He moves ahead, angles to the right to ease his climb up a steep dunelet, detours around a pebble, stops for a moment to exchange information with a compatriot. Thus he makes his weaving, halting way back to his home. So as not to anthropomorphize about his purposes, I sketch the path on a piece of paper. It is a sequence of irregular, angular segments—not quite a random walk, for it has an underlying sense of direction, of aiming toward a goal.

I show the unlabeled sketch to a friend. Whose path is it? An expert skier, perhaps, slaloming down a steep and somewhat rocky slope. Or a sloop, beating upwind in a channel dotted with islands or shoals. Perhaps it is a path in a more abstract space: the course of search of a student seeking the proof of a theorem in geometry.

Whoever made the path, and in whatever space, why is it not straight; why does it not aim directly from its starting point to its goal? In the case of the ant (and, for that matter, the others), we know the answer. He has a general sense of where home lies, but he cannot foresee all the obstacles between. He must adapt his course repeatedly to the difficulties he encounters, and often detour uncrossable barriers. His horizons are very close, so that he deals with each obstacle as he comes to it; he probes for ways around or over it, without much thought for future obstacles. It is easy to trap him into deep detours.

Viewed as a geometric figure, the ant's path is irregular, complex, hard to describe. But its complexity is really a complexity in the surface of the beach, not a complexity in the ant. On that same beach, another small creature, with a home at the same place as the ant, might well follow a very similar path.

Some years ago, Grey Walter built an electromechanical "turtle" capable of exploring a surface and of periodically seeking its nest, where its batteries were recharged. More recently, goal-seeking automata have been under construction in several laboratories, including Professor Marvin Minsky's in Cambridge, Massachusetts. Suppose we undertook to design such an automaton with the approximate dimensions of an ant, similar means of locomotion, and comparable sensory acuity. Suppose we provided it with a few simple adaptive capabilities: when faced with a steep slope, try climbing it obliquely; when faced with an insuperable obstacle, try detouring; and so on. (Except for problems of miniaturization of components, the present state of the art would surely support such a design.) How different would its behavior be from the behavior of the ant?

These speculations suggest a hypothesis, one that could as well have been derived as corollary from our previous discussion of artificial objects:

An ant, viewed as a behaving system, is quite simple. The apparent complexity of its behavior over time is largely a reflection of the complexity of the environment in which it finds itself.

We may find this hypothesis initially plausible or implausible. It is an empirical hypothesis, to be tested by seeing whether attributing quite simple properties to the ant's adaptive system will permit us to account for its behavior in the given or similar environments. For the reasons developed at length in my previous chapter, the truth or falsity of the hypothesis should be independent of whether ants, viewed more microscopically, are simple or complex systems. At the level of cells or molecules, ants are demonstrably complex; but these microscopic details of the inner environment may be largely irrelevant to the ant's behavior in relation to the outer environment. That is why an automaton, though completely different at the microscopic level, might nevertheless simulate the ant's gross behavior.

In this chapter, I should like to explore this hypothesis, but with the word "man" substituted for "ant."

A man, viewed as a behaving system, is quite simple. The apparent complexity of his behavior over time is largely a reflection of the complexity of the environment in which he finds himself.

Now I should like to hedge my bets a little. Instead of trying to consider the "whole man," fully equipped with glands and viscera, I should like to limit the discussion to Homo sapiens, "thinking man." I myself believe that the hypothesis holds even for the whole man, but it may be more prudent to divide the difficulties at the outset, and analyze only cognition rather than behavior in general.[1]

The reasons for assigning some a priori probability to the hypothesis have already been set forth in the last chapter. A thinking human being is an adaptive system; his goals define the interface between his inner and outer environments. To the extent that he is effectively adaptive, his behavior will reflect characteristics largely of the outer environment (in the light of his goals) and will reveal only

[1] I have sketched an extension of this hypothesis to phenomena of emotion and motivation in "Motivational and Emotional Controls of Cognition," *Psychological Review*, *74*: 29–39 (1967), and to certain aspects of perception in "An Information-Processing Explanation of Some Perceptual Phenomena," *British Journal of Psychology*, *58*: 1–12 (1967). Both of these areas would seem to require, however, more specification of physiological structure than is involved in the cognitive phenomena considered in this volume.

a few limiting properties of his inner environment—of the physiological machinery that enables him to think.

I do not intend to repeat this theoretical argument at length, but rather I want to seek empirical verification for it in the realm of human thought processes. Specifically, I should like to point to evidence that there are only a few "intrinsic" characteristics of the inner environment of thinking man that limit the adaptation of his thought to the shape of the problem environment. All else in his thinking and problem-solving behavior is artificial—is learned and is subject to improvement through the invention of improved designs.

PSYCHOLOGY AS A SCIENCE OF THE ARTIFICIAL

Problem solving is often described as a search through a vast maze of possibilities, a maze that describes the environment. Successful problem solving involves searching the maze selectively and reducing it to manageable proportions. Let us take, by way of specific example, a puzzle of the kind known as cryptarithmetic problems:[2]

$$\begin{array}{r} DONALD \\ +GERALD \\ \hline ROBERT \end{array} \qquad D=5$$

The task is to replace the letters in this array by numerals, from zero through nine, so that all instances of the same letter are replaced by the same numeral, different letters are replaced by different numerals, and the resulting numerical array is a correctly worked out problem in arithmetic. As an additional hint for this particular problem, the letter D is to be replaced by the numeral 5.

[2] The cryptarithmetic task was first used for research on problem solving by F. Bartlett in his *Thinking* (New York: Basic Books, 1958). In the present account, I have drawn on his work, on a report by my colleague Allen Newell, *Studies in Problem Solving: Subject 3 on the Crypt-airthmetic Task DONALD + GERALD = ROBERT* (Pittsburgh: Carnegie Institute of Technology, July 1966, mimeographed), and on a forthcoming analysis by Newell and myself, of a number of problem-solving protocols on this task.

One way of viewing this task is to consider all the 10!, ten factorial, ways in which ten numerals can be assigned to ten letters. The number 10! is not so large as to strike awe in the heart of a modern computer; it is only a little more than 3 million (3,628,800, to be exact). A program designed to generate all possible assignments systematically, and requiring a tenth of a second to generate and test each, would require at most about ten hours to do the job. (With the cue $D = 5$, only an hour would be needed.) I haven't written the program, but I suspect that a tenth of a second is far longer than a large computer would need to examine each possibility.

There is no evidence that a human being could do this. It might take him as long as a minute to generate and test each assignment, and he would have great difficulty in keeping track of where he was and what assignments he had already tried. He could use paper and pencil to assist him on the latter score, but that would slow him down even more. The task, performed in this way, might call for several man-years of work—I assume a forty-hour week.

Notice that in excluding exhaustive, systematic search as a possible way for a human to solve the problem, we are making only very gross assumptions about human capabilities. We are assuming that simple arithmetic operations take times that are of the order of seconds, that the operations are essentially executed serially, rather than in parallel, and that large amounts of memory are not available, in which new information can be stored at split-second speeds. These assumptions say something, but not very much, about the physiology of the human central nervous system. For example, modifying the brain by incorporating in it a new subsystem with all the properties of a desk calculator would be a quite remarkable feat of brain surgery—or evolution. But even such a radical alteration would change the relevant assumptions only slightly for purposes of explaining or predicting behavior in this problem environment.

Human beings do frequently solve the *DONALD* + *GERALD* = *ROBERT* problem. How do they do it? What are the alternative ways of representing the environment and conducting the search?

Search Strategies

One way to cut down the search drastically is to make the assignments systematically, as before, but to assign numerals to the letters one by one so that inconsistencies can be detected before an assignment is complete, and hence whole classes of possible assignments can be ruled out at one step. Let me illustrate how this works.

Suppose we start from the right, trying assignments successively for the letters *D*, *T*, *L*, *R*, *A*, *E*, *N*, *B*, *O*, and *G*, and substituting numerals in the order 1, 2, 3, 4, 5, 6, 7, 8, 9, 0. We already know that $D = 5$, so we strike 5 from the list of available numerals. We now try $T = 1$. Checking in the right-hand column, we detect a contradiction, for $D + D = T + c$, where c is 10 or 0. Hence, since $[D = 5, T = 1]$ is not feasible, we can rule out all the remaining 8! assignments of the eight remaining numerals to the eight remaining letters. In the same way, all possible assignments for *T*, except $T = 0$, can be ruled out without considering the assignments for the remaining letters.

The scheme can be improved further by the expedient of calculating directly, by addition, what assignment should be made to the sum of a column whenever the two addends are known. With this improvement, we shall not need to search for the assignment for *T*, for $T = 0$ can be inferred directly from $D = 5$. Using this scheme, the $DONALD + GERALD = ROBERT$ problem can be solved quite readily, with paper and pencil. Ten minutes should suffice. Figure 3 shows the search tree, in slightly simplified form. Each branch is carried to the point where a contradiction is detected. For example, after the assignments $[D = 5, T = 0]$, the assignment $L = 1$ leads to the inference $R = 3$, which yields a contradiction since, from the left-hand column of the problem array $R = 3$ would imply that *G* is negative.

Figure 3 is oversimplified in one respect. Each of the branches that terminates with a contradiction after assignment of a value to *E* should actually be branched one step further. For the contradiction in these cases arises from observing that no assignment for the letter *O* is now consistent. In each case, four assignments must be

```
D = 5  T = 0  L = 1  R = 3  G < 0  □
              L = 2  R = 5  G = 0  □
              L = 3  R = 7  A = 1  E = 2  □
                            A = 2  E = 4  □
                            A = 4  E = 8  □
                            A = 6  E = 2  □
                            A = 8  E = 6  □
                            A = 9  E = 8  □
              L = 4  R = 9  A = 1  E = 2  □
                            A = 2  E = 4  □
                            A = 3  E = 6  □
                            A = 6  E = 2  □
                            A = 7  E = 4  □
                            A = 8  E = 6  □
              L = 6  R = 3  G < 0  □
              L = 7  R = 5  □
              L = 8  R = 7  A = 1  E = 3  □
                            A = 2  E = 5  □
                            A = 3  E = 7  □
                            A = 4  E = 9  N = 1  B = 8  □
                                          N = 2  B = 9  □
                                          N = 3  G = 0  □
                                          N = 6  O = 2  G = 1
```

Figure 3. Possible Search Tree for
DONALD + GERALD = ROBERT

examined to determine this. Thus, the full search tree would have 68 branches—still a far cry from 10! or even 9!

An enormous space has been cut down to a quite small space by some relatively small departures from systematic, exhaustive search. It must be confessed that the departures are not all as simple as I have made them appear. One step in the proposed scheme requires finding the contradictions that are implied by an assignment. This means, of course, the "relatively direct" contradictions, for if we had a rapid process capable of detecting *all* inconsistent implications, direct or indirect, it would find the problem solution almost at once. For, in this problem, any set of assignments other than the single correct one implies a contradiction.

What is meant by searching for direct contradictions is something like this: After a new assignment has been made, those columns are examined where the newly substituted letter occurs. Each such column is solved, if possible, for a still-unassigned letter, and the solution checked to see whether this numeral remains unassigned. If not, there is a contradiction.

In place of brute-force search, we have now substituted a combined system of search and "reason." Can we carry this process further; can we eliminate substantially all trial-and-error search from the solution method? It turns out that we can for this problem, although not for all cryptarithmetic problems.[3]

The basic idea that permits us to eliminate most trial-and-error search in solving the problem before us is to depart from the systematic right-to-left assignment of numerals. Instead, we search for columns of the problem array that are sufficiently determinate to allow us to make new assignments, or at least new inferences about the properties of assignments.

Let me go through the process briefly. From $D = 5$, we immediately infer $T = 0$, as before. We also infer that 1 is carried into the second column, hence that $R = 2L + 1$ is odd. On the extreme left, from $D = 5$, we infer that R is greater than 5 (for $R = 5 + G$). Putting together these two inferences, we have $R = 7$ or $R = 9$, but we do not try these assignments. Now we discover that the second column from the left has the peculiar structure $O + E = O$ —a number plus another equals itself (apart from what is carried into or out of the column). Mathematical knowledge, or experiment, tells us that this can be true only if $E = 0$ or $E = 9$. Since we already have $T = 0$, it follows that $E = 9$. This eliminates one of the alternatives for R, so $R = 7$.

Since $E = 9$, it follows that $A = 4$, and there must be a one carried into the third column from the right; hence $2L + 1 = 17$, or $L = 8$. All that remains now is to assign 1, 2, 3, and 6 in some order to N, B, O, and G. We get $G = 1$ by observing that, for any assignment of O, there is a number carried into the leftmost column. We are now left with only $3! = 6$ possibilities, which we may be willing to eliminate by trial and error: $N = 6$, $B = 3$, and therefore $O = 2$.

We have traced a solution path through the problem maze on three different assumptions about the search strategy. The more sophisticated, in a certain sense, that strategy became, the less search was required. But it is important to notice that, once the strategy was selected, the course of the search depended only on the structure of the

[3] For example, the method to be described does not eliminate as much search from the cryptarithmetic problem *CROSS* + *ROADS* = *DANGER*.

problem, not on any characteristics of the problem solver. By watching a man, or an automaton, perform in this problem environment, what could we learn about him? We might well be able to infer what strategy he followed. By mistakes he made, and his success in recovering from them, we might be able to detect certain limits of the capacity or accuracy of his memory and his elementary processes. We might learn something about the speed of these processes. Under favorable circumstances, we might be able to learn which among the thinkable strategies he was able actually to acquire, and under what circumstances he was likely to acquire them. We should certainly be unlikely to learn anything specific about the neurological characteristics of his central nervous system, nor would the specifics of that system be relevant to his behavior, beyond placing bounds on the possible.

The Limits on Performance

Let us undertake to state in positive fashion just what we think these bounds and limits are, as revealed by behavior in problem situations like this one. In doing so, we shall draw upon both experimental evidence and evidence derived from computer simulations of human performance. The evidence refers to a variety of cognitive tasks, ranging from relatively complex ones (cryptarithmetic, chess, theorem proving), through an intermediate one (concept attainment), to simple ones that have been favorities of the psychological laboratory (rote verbal learning, short-term memory span). It is important that with this great variety of performance only a small number of limits on the adaptability of the inner system reveal themselves—and these are essentially the same limits over all the tasks. Thus, the statement of what these limits are purports to provide a single, consistent explanation of human performance over this whole range of heterogeneous task environments.

Limits on Speed of Concept Attainment

Extensive psychological research has been carried out on

concept attainment within the following general paradigm.[4] The stimuli are a set of cards bearing simple geometric designs that vary, from card to card, along a number of dimensions: shape (square, triangle, circle), color, size, position of figure on card, and so on. A "concept" is defined extensionally by some set of cards—the cards that are instances of that concept. The concept is defined intensionally by a property that all the instances have in common but that is not possessed by any of the remaining cards. Examples of concepts are "yellow" or "square" (simple concepts), "green triangle" or "large, red" (conjunctive concepts), "small or yellow" (disjunctive concept), and so on.

In our discussion here I shall refer to experiments using an *N*-dimensional stimulus, with two possible values on each dimension, and with a single relevant dimension (simple concepts). On each trial, an instance (positive or negative) is presented to the subject; he responds "Positive" or "Negative"; and he is reinforced by "Right" or "Wrong," as the case may be. In typical experiments of this kind, the subject's behavior is reported in terms of number of trials or number of erroneous responses before he attains an error-free performance. Some, but not all, experiments ask the subject also to report periodically the intensional concept (if any) he is using as a basis for his responses.

This situation is so simple that, as in the cryptarithmetic problem, we can estimate a priori how many trials, on the average, a subject should need to discover the intended concept provided that he used the most efficient discovery strategy. On each trial, regardless of what response the subject makes, he can determine from the experimenter's reinforcement whether the stimulus was actually an instance of the concept or not. If it was an instance, he knows that one of the attribute values of the stimulus—its color, size, shape, for example—defines the concept. If it was not an

[4] This account of concept attainment is based on the paper with my colleague Lee Gregg, "Process Models and Stochastic Theories of Simple Concept Formation," *Journal of Mathematical Psychology*, *4*: 246–276 (June 1967). See also A. Newell and H. A. Simon, "Overview: Memory and Process in Concept Formation," Chapter 11 in B. Kleinmuntz (ed.), *Concepts and the Structure of Memory* (New York: Wiley, 1967), pp. 241–262.

instance, he knows that the *complement* of one of its attribute values defines the concept. In either case, each trial rules out half of the possible simple concepts; and in a random sequence of stimili, each new stimulus rules out, on the average, approximately half of the concepts not previously eliminated. Hence, the average number of trials required to find the right concept will vary with the logarithm of the number of dimensions in the stimulus.

If sufficient time were allowed for each trial (a minute, say, to be generous), and if the subject were provided with paper and pencil, any subject of normal intelligence could be taught to follow this most efficient strategy, and would do so without much difficulty. As these experiments are actually run, subjects are not instructed in an efficient strategy, are not provided with paper and pencil, and take only a short time—typically four seconds, say—to respond to each successive stimulus. They also use many more trials to discover the correct concept than the number calculated from the efficient strategy. Although the experiment has not, to my knowledge, been run, it is fairly certain that, even with training, a subject who was required to respond in four seconds and not allowed paper and pencil would be unable to apply the efficient strategy.

What do these experiments tell us about human thinking? First, they tell us that human beings do not always discover for themselves clever strategies that they could readily be taught (watching a chess master play a duffer should also convince us of that). This is hardly a very startling conclusion, although it may be an instructive one. I shall return to it in a moment.

Second, the experiments tell us that human beings do not have sufficient means for storing information in memory to enable them to apply the efficient strategy unless the presentation of stimuli is greatly slowed down, or unless the subjects are permitted external memory aids, or both. Since we know from other evidence that human beings have virtually unlimited semipermanent storage (as indicated by their ability to continue to store odd facts in memory over most of a lifetime), the bottleneck in the experiment must lie in the small amount of rapid-access storage (so-called short-term memory) available and the time required to move items from the limited short-term store to the large-scale long-term store.[5]

From evidence obtained in other experiments, it has been estimated that only some seven items (or perhaps as few as two) can be held in the fast, short-term memory and that perhaps as many as five seconds are required to transfer an item from the short-term to the long-term store. To make these statements operational, we shall have to be more precise, presently, about the meaning of "item." For the moment, let us assume that a simple concept is an item.

Even without paper and pencil, a subject might be expected to apply the efficient strategy if: (1) he was instructed in efficient strategy; and (2) he was allowed twenty or thirty seconds to respond to and process the stimulus on each trial. Since I have not run the experiment, this assertion stands as a prediction by which the theory may be tested.

Again, the outcome may appear obvious to you, if not trivial. If so, I remind you that it is obvious only if you accept my general hypothesis: that in large part human goal-directed behavior simply reflects the shape of the environment in which it takes place; only a gross knowledge of the characteristics of the human information-processing system is needed in order to predict it. In this experiment, the relevant characteristics appear to be: (1) the capacity of short-term memory, measured in terms of number of items (or "chunks," as I shall call them); (2) the time required to fixate an item, or chunk, in long-term memory. In the next section I shall inquire as to how consistent these characteristics appear to be over a range of task environments. Before I do so, I want to make a concluding comment about subjects' knowledge of strategies and the effects of training subjects.

That strategies can be learned is hardly a surprising fact, nor that learned strategies can vastly alter performance and enhance its effectiveness. All educational institutions

[5] The monograph by J. S. Bruner, J. J. Goodnow, and G. A. Austin, *A Study of Thinking* (New York: Wiley, 1956) was perhaps the first work to emphasize the role of short-term memory limits (their term was "cognitive strain") in performance on concept-attainment tasks. That work also provided rather definite descriptions of some of the subjects' strategies.

are erected on these premises. Their full implication has not always been drawn by psychologists who conduct experiments in cognition. Insofar as behavior is a function of learned technique rather than "innate" characteristics of the human information-processing system, our knowledge of behavior must be regarded as sociological in nature rather than psychological—that is, as revealing what human beings in fact learn when they grow up in a particular social environment. When and how they learn particular things may be a difficult question, but we must not confuse learned strategies with built-in properties of the underlying biological system.

The data that have been gathered, by Bartlett and in our own laboratory, on the cryptarithmetic task illustrate the same point. Different subjects do, indeed, apply different strategies in that task—both the whole range of strategies I sketched in the previous section and others as well. How they learned these, or how they discover them while performing the task, we do not know, although we know that the sophistication of the strategy varies directly with a subject's previous exposure to and comfort with mathematics. But, apart from the strategies, the only human characteristic that exhibits itself strongly in the cryptarithmetic task is the limited size of short-term memory. Most of the difficulties the subjects have in executing the more combinatorial strategies (and perhaps their general aversion to these strategies also) stem from the stress that such strategies place on short-term memory. Subjects get into trouble simply because they forget where they are, what assignments they have made previously, and what assumptions are implicit in assignments they have made conditionally. All of these difficulties would necessarily arise in a processor that could hold only a few chunks in short-term memory and that required more time than was available to transfer them to long-term memory.

THE PARAMETERS OF MEMORY: FIVE SECONDS PER CHUNK

If a few parameters of the sort we have been discussing are the main limits of the inner system that reveal themselves in human cognitive behavior, then it becomes an important task for experimental psychology to estimate the values of these parameters and to determine how variable or constant they are among different subjects and over different tasks.

Apart from some areas of sensory psychology, the typical experimental paradigms in psychology are concerned with hypothesis testing rather than parameter estimating. In the reports of experiments one can find many assertions that a particular parameter value is—or is not—"significantly different" from another but very little comment on the values themselves. As a matter of fact, the pernicious practice is sometimes followed of reporting significance levels, or results of the analysis of variance, without reporting at all the numerical values of the parameters that underlie these inferences.

While I am objecting to publication practices in experimental psychology, I shall add another complaint. Typically, little care is taken in choosing measures of behavior that are the most relevant to theory. Thus, in learning experiments, "rate of learning" is reported, almost indifferently, in terms of "number of *trials* to criterion," "total number of *errors*," "total *time* to criterion," and perhaps other measures as well. Specifically, the practice of reporting learning rates in terms of trials rather than time, prevalent through the first half of this century, and almost up to the present time, not only hid from view the remarkable constancy of the parameter I am about to discuss, but also led to much meaningless dispute over "one-trial" versus "incremental" learning.[6]

Ebbinghaus knew better. In his classic experiments on learning nonsense syllables, with himself as subject, he recorded both the number of repetitions and the amount of

[6] The evidence of the constancy of the fixation parameter is reviewed in L. W. Gregg and H. A. Simon, "An Information-Processing Explanation of One-Trial and Incremental Learning," *Journal of Verbal Learning and Verbal Behavior*, *6*: 780–787 (1967); H. A. Simon and E. A. Feigenbaum, "An Information-Processing Theory in Verbal Learning," *ibid.*, *3*: 385–396 (1964); Feigenbaum and Simon, "A Theory of the Serial Position Effect," *British Journal of Psychology*, *53*: 307–320 (1962); E. A. Feigenbaum, "An Information-Processing Theory of Verbal Learning," unpublished doctoral dissertation, Pittsburgh: Carnegie Institute of Technology, 1959; and references cited therein.

time required to learn sequences of syllables of different length. If you take the trouble to calculate it, you find that the *time per syllable* in his experiments works out to about ten to twelve seconds.[7]

I see no point in computing the figure to two decimal places—or even to one. The constancy here is a constancy to an order of magnitude, or perhaps to a factor of two—more nearly comparable to the constancy of the daily temperature, which in most places stays between 263° and 333° Kelvin, than to the constancy of the speed of light. There is no reason to be disdainful of a constancy to a factor of two. Newton's original estimates of the speed of sound contained a fudge factor of 30 per cent (eliminated only a hundred years later), and today some of the newer physical "constants" for elementary particles are even more vague. Beneath any approximate, even very rough, constancy, we can usually expect to find a genuine parameter whose value can be defined accurately once we know what conditions we must control during measurement.

If the constancy simply reflected a parameter of Ebbinghaus—albeit one that held steady over several years—it would be more interesting to biography than psychology. But that is not the case. When we examine some of the Hull-Hovland experiments of the 1930's, as reported, for example, in Carl Hovland's chapter in S. S. Stevens' *Handbook*, we find again (after we calculate them, for trials are reported instead of times) times in the neighborhood of ten or fifteen seconds for college sophomores to fixate nonsense syllables of low meaningfulness by the serial anticipation method. When the drum speed increases (say from four seconds per syllable to two seconds per syllable), the number of trials to criterion increases proportionately, but the total learning time remains essentially constant.

There is a great deal of gold in these hills. If past nonsense-syllable experiments are re-examined from this point of view, many are revealed where the basic learning parameter is in the neighborhood of fifteen seconds per

[7] Herman Ebbinghaus, *Memory* (New York: Dover Publications, 1964), translated from the German edition of 1885, especially pp. 35–36, 40, 51.

syllable. You can make the calculation yourself from the experiments reported, for example, in J. A. McGeoch's *Psychology of Human Learning*. B. R. Bugelski, however, seems to have been the first to make this parameter constancy a matter of public record and to have run experiments with the direct aim of establishing it.[8]

I have tried not to exaggerate how constant is "constant." On the other hand, efforts to purify the parameter measurement have hardly begun. We do know about several variables that have a major effect on the value, and we have a theoretical explanation of these effects that, thus far, has held up well.

We know that meaningfulness is a variable of great importance. Nonsense syllables of high association value and unrelated one-syllable words are learned in about one third the time required for nonsense syllables of low association value. Continuous prose is learned in about one third the time, per word, required for sequences of unrelated words. (We can get the latter figure also from Ebbinhaus' experiments in memorizing *Don Juan*. The times *per symbol* are roughtly 10 per cent of the corresponding times for nonsense syllables.)

We know that similarity—particularly similarity among stimuli—has an effect on the fixation parameter somewhat less than the effect of meaningfulness, and we can also estimate its magnitude on theoretical grounds.

The theory that has been most successful in explaining these and other phenomena reported in the literature on rote verbal learning is an information-processing theory, programmed as a computer simulation of human behavior, dubbed EPAM. Since EPAM has been reported at length in the literature, I shall not discuss it here, except for one point that is relevant to our analysis. The EPAM theory gives us a basis for understanding what a "chunk" is. A chunk is a maximal familiar substructure of the stimulus. Thus, a nonsense syllable like "QUV" consists of the chunks "Q," "U," "V"; but the word "CAT" consists of a single chunk, since it is a highly familiar unit. EPAM postulates constancy in the time required to

[8] B. R. Bugelski, "Presentation Time, Total Time, and Mediation in Paired-Associate Learning," *Journal of Experimental Psychology*, *63*: 409–412 (1962).

fixate a chunk. Empirically, the constant appears to be about five seconds per chunk, or perhaps a little less. Virtually all the quantitative predictions that EPAM makes about the effects of meaningfulness, familiarity, and similarity upon learning speed follow from this conception of the chunk and of the constancy of the time required to fixate a single chunk.[9]

THE PARAMETERS OF MEMORY— SEVEN CHUNKS, OR IS IT TWO?

The second limiting property of the inner system that shows up again and again in learning and problem-solving experiments is the amount of information that can be held in short-term memory. Here, again, the relevant unit appears to be the chunk, where this term has exactly the same meaning as in the definition of the fixation constant.

Attention was attracted to this parameter, known previously from digit-span, numerosity-judging, and discrimination tasks, by George Miller's justly celebrated paper on "The Magical Number Seven, Plus or Minus Two."[10] It is no longer as plausible as it was when he wrote his paper that a single parameter is involved in the three kinds of task, rather than three different parameters; we shall consider here only tasks of the digit-span variety. Nor is it entirely clear today whether the correct value of the parameter is seven or two—rather too large a range for comfort.

The facts that appear to emerge from recent experiments on short-term memory are these. If a subject is to read a string of digits or letters and is asked simply to repeat them back, he can generally perform correctly on strings up to seven or even ten items in length. If almost any other

[9] Most of the experiments against which EPAM has been tested employ recall tasks rather than simply recognition. EPAM contains two subprocesses that are implicated in fixation: a process of learning to differentiate, and an image-storing process. Since recognition involves largely the former rather than the latter, it is possible that study of recognition tasks may reveal a fixation parameter smaller than five seconds per chunk. Indeed, experiments reported to me by Walter Reitman and by Dr. Mary Potter (private communications) suggest that the time required to store the fact of having just seen familiar objects or pictures of them may be on the order of a second or two.

[10] *Pyschological Review*, *63*: 81–97 (1956).

task, however simple, is interposed between his hearing the items and his repeating them, the number he can retain drops to two. From its familiarity in daily life, we could dub these numbers the "telephone directory constants." We can generally retain seven numbers from directory to phone if we are not interrupted in any way—not even by our own thoughts.

Where experiments appear to show that more than two chunks are retained across an interruption, the phenomena can almost always be explained parsimoniously by mechanisms we have already discussed in the previous section. In some of these experiments, the explanation—as already pointed out by Miller—is that the subject recodes the stimulus into a smaller number of chunks before storing it in short-term memory. If ten items can be recorded as two chunks, then ten items can be retained. In the other experiments where "too much" appears to be retained in short-term memory, the times allowed the subjects permit them, in fact, to fixate the excess of items in long-term memory.

I shall cite just two examples from the literature. N. C. Waugh and D. A. Norman report experiments, their own and others',[11] which show that only the first two of a sequence of items is retained reliably across interruption but that there is some residual retention of the remaining items. Computation of the fixation times available to the subjects in these experiments shows that a transfer rate to long-term memory of one chunk per five seconds would explain most of the residuals. (This explanation is entirely consistent with the theoretical model that Waugh and Norman themselves propose.)

Recently, Roger Shepard has reported that subjects shown a very long sequence of photographs—mostly landscapes—can remember which of these they have seen (when asked to choose from a large set) with high reliability.[12] When we note that the task is a recognition task, requiring storage only of differentiating cues, and that the average time per item was about six seconds, the

[11] N. C. Waugh and D. A. Norman, "Primary Memory," *Psychological Review*, *72*: 89–104 (1965).

[12] Roger N. Shepard, "Recognition Memory for Words, Sentences, and Pictures," *Journal of Verbal Learning and Verbal Behavior*, *6*: 156–163 (1957).

phenomena become entirely understandable—indeed, predictable—within the framework of the theory that we are proposing.

The Organization of Memory

I have by no means exhausted the list of experiments I could cite in support of the fixation parameter and the short-term capacity parameter and in support of the hypothesis that these parameters are the principal, and almost only, characteristics of the information-processing system that are revealed, or could be revealed, by these standard psychological experiments.

This does not imply that there are not other parameters, and that we cannot find experiments in which they are revealed and from which they can be estimated. What it does imply is that we should not look for great complexity in the laws governing human behavior, in situations where the behavior is truly simple and only its environment is complex.

In our laboratory, we have found that mental arithmetic tasks, for instance, provide a useful environment for teasing out other possible parameters. Work that Dansereau has been carrying forward shows that the times required for elementary arithmetic operations and for fixation of intermediate results account for only part—perhaps one half—of the total time for performing mental multiplications of four digits by two. Much of the remaining time appears to be devoted to retrieving numbers from the memory where they have been temporarily fixated, and "placing" them in position in short-term memory where they can be operated upon. We hope, through Dansereau's work, to arrive at estimates for these new parameters and at some understanding of the processes that underlie them.[13]

Stimulus Chunking

I should like now to point to another kind of characteristic

[13] The first results of this work, relating to the over-all time requirements of the mental arithmetic tasks, are reported in Donald F. Dansereau and Lee W. Gregg, "An Information Processing Analysis of Mental Multiplication," *Psychonomic Science*, *6*: 71–72 (1966).

of the inner system—more "structural," and also less quantitative—that is revealed in certain experiments. Memory is generally conceived to be organized in an "associative" fashion, but it is less clear just what that term is supposed to mean. One thing it means is revealed by McLean and Gregg. They gave subjects lists to learn—specifically, 24 letters of the alphabet in scrambled order. They encouraged, or induced, chunking of the lists by presenting the letters either one at a time, or three, four, six, or eight on a single card. In all of the grouped conditions, subjects learned in about half the time required in the one-at-a-time condition.[14]

McLean and Gregg also sought to ascertain whether the learned sequence was stored in memory as a single long list or as a hierarchized list of chunks, each of which was a shorter list. They determined this by measuring how subjects grouped items temporally when they recited the list, and especially when they recited it backwards. The results were clear: the alphabets were stored as sequences of short subsequences; the subsequences tended to correspond to chunks presented by the experimenter, or sublengths of those chunks; left to his own devices, the subject tended to prefer chunks of three or four letters. (Recall the role of chunks of this length in the experiments on effects of meaningfulness in rote learning.)

Visual Memory

The materials in the McLean–Gregg experiments were strings of symbols. We might raise similar questions regarding the form of storage of information about two-dimensional visual stimuli.[15] In what sense do memory and thinking represent the visual characteristics of stimuli? I do not wish to revive the debate on "imageless

[14] R. S. McLean and L. W. Gregg, "Effects of Induced Chunking on Temporal Aspects of Serial Recitation," *Journal of Experimental Psychology*, *74*: 455–459 (1967).

[15] The letters in the stimuli of the McLean–Gregg experiment are, of course, also two-dimensional visual stimuli. Since they are familiar chunks, however, and can be immediately recognized and recoded, there is no reason to suppose that their two-dimensional character plays any role in the subject's behavior in the experiment. Again this is "obvious," but only if we already have a general theory of how stimuli are processed "inside."

thought"—certainly not in the original form that debate took. But perhaps the issue can now be made more operational than it was at the turn of the century.

As I enter into this dangerous ground, I am comforted by the thought that even the most fervent opponents of mentalism have preceded me. I quote, for example, from B. F. Skinner's *Science and Human Behavior* (1953, p. 266):

> A man may see or hear "stimuli which are not present" on the pattern of the conditioned reflexes: he may see X, not only when X is present, but when any stimulus which has frequently accompanied X is present. The dinner bell not only makes our mouth water, it makes us see food.

I do not know exactly what Professor Skinner means by "seeing food," but his statement gives me courage to say what an information-processing theory might mean by it. I shall describe in a simplified form one kind of experiment that has been used to throw light on the question. Suppose we allow a subject to memorize the following visual stimulus—a magic square:

4 9 2

3 5 7

8 1 6

Now we remove the stimulus and ask the subject a series of questions about it, timing his answers. What numeral lies to the right of 3, to the right of 1? What numeral lies just below 5? What numeral is diagonally above and to the right of 3? The questions are not all of the same difficulty—in fact, I have arranged them in order of increasing difficulty, and would expect a subject to take substantially longer to answer the last question than the first.

Why should this be? If the image stored in memory were isomorphic to a photograph of the stimulus, we should expect no large differences in the times required to answer the different questions. We must conclude that the stored image is organized quite differently from a photograph. An alternative hypothesis is that it is a list structure—a hypothesis that is consistent, for example, with the data from the McLean–Gregg experiment and that is much in the spirit of information-processing models of cognition.

For example, if what was stored were a list of lists: "TOP," "MIDDLE," "BOTTOM," where "TOP" is 4–9–2, "MIDDLE" is 3–5–7, and "BOTTOM" is 8–1–6; the empirical results would be easy to understand. The question "What numeral lies to the right of 3?" is answered by searching down lists. The question "What numeral lies just below 5?" is answered, on the other hand, by matching two lists, item by item—a far more complex process than the previous one.

There is no doubt, of course, that a subject could *learn* the up-down relations or the diagonal relations as well as the left-right relations. An EPAM-like theory would predict that it would take the subject about twice as long to learn both left-right and up-down relations as the former alone. This hypothesis can be easily tested, but, to the best of my knowledge, it has not been.

Evidence about the nature of the storage of "visual" images, pointing in the same direction as the example I have just given, is provided by A. de Groot's well-known experiments on chess perception.[16] De Groot put chess positions—taken from actual games—before subjects for, say, five seconds; then he removed the positions and asked the subjects to reconstruct them. Chess grandmasters and masters could reconstruct the positions (with perhaps 20 to 24 pieces on the board) almost without error, while duffers were able to locate hardly any of the pieces correctly, and the performance of players of intermediate skill fell somewhere between masters and duffers. But the remarkable fact was that, when masters and grandmasters were shown other chessboards with the same numbers of pieces *arranged at random*, their abilities to reconstruct the boards were no better than the duffers' with the boards from actual games, while the duffers performed as well or poorly as they had before.

What conclusion shall we draw from the experiment? The data are inconsistent with the hypothesis that the

[16] Adriaan D. de Groot, "Perception and Memory versus Thought: Some Old Ideas and Recent Findings," in B. Kleinmuntz (ed.), *Problem Solving* (New York: Wiley, 1966), pp. 19–50.

chess masters have some special gift of visual imagery—or else why the deterioration of their performance? What the data suggest strongly is that the information about the board is stored in the form of *relations* among the pieces, rather than a "television scan" of the 64 squares. It is inconsistent with the parameters proposed earlier—seven chunks in short-term memory and five seconds to fixate a chunk—to suppose that anyone, even a grandmaster, can store 64 pieces of information (or 24) in ten seconds. It is quite plausible that he can store (in short-term and long-term memory) information about enough relations (supposing each one to be a familiar chunk) to permit him to reproduce the board of Figure 4:

1. Black has castled on the K's side, with a fianchettoed K's bishop defending the K's Knight.
2. White has castled on the Q's side, with his Queen standing just before his King.
3. A Black pawn on his K5 and a White pawn on his Q5 are attacked and defended by their respective K's and Q's Knights, the White Queen also attacking the Black pawn on the diagonal.
4. White's Q-Bishop attacks the Knight from KN5.
5. The Black Queen attacks the White K's position from her QN3.
6. A Black pawn stands on its QB4.
7. A White pawn on K3 blocks that advance of the opposing Black pawn.
8. Each side has lost a pawn and a Knight.
9. White's K-Bishop stands on K2.

Pieces not mentioned are assumed to be in their starting positions. Since some of the relations as listed are complex, I shall have to provide reasons for considering them unitary "chunks." I think most strong chess players would regard them as such. Incidentally, I wrote down these relations from my own memory of the position, in the order in which they occurred to me. Eye-movement data for an expert chess player looking at this position tend to support this analysis of how the relations are analyzed and stored.[17] The eye-movement data exhibit especially clearly the relations 3 and 5.

[17] O. K. Tikhomirov and E. D. Poznyanskaya, "An Investigation of Visual Search as a Means of Analysing Heuristics," English

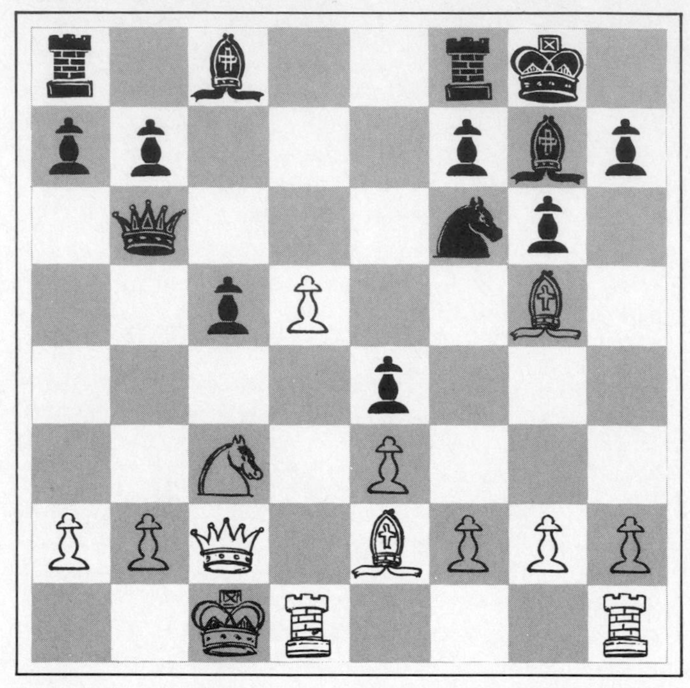

Figure 4. Chess Position Used in Memory Experiment

The implication of this discussion of visual memory for my main theme is that many of the phenomena of visualization do not depend in any detailed way upon underlying neurology but can be explained and predicted on the basis of quite general and abstract features of the organization of memory—features which are essentially the same ones that were postulated in order to build information-processing theories of rote learning and of concept attainment phenomena.

Specifically, we are led to the hypothesis that memory is an organization of list structures (lists whose components can also be lists), which include descriptive components (two-termed relations) and short (three-element or four-element) component lists. A memory having this form of organization appears to have the right properties to explain phenomena relating to storage of visual and auditory stimuli as well as "symbolic" stimuli.

---

translation from *Voprosy psikhologii*, 1966, Vol. 12, in *Soviet Psychology*, Vol. 2, No. 2 (Winter 1966–1967), pp. 3–15.

Processing Natural Language

A theory of human thinking cannot and should not avoid reference to that most characteristic cognitive skill of human beings—the use of language. How does language fit into the general picture of cognitive processes that I have been sketching and into my general thesis that psychology is a science of the artificial?

Historically, the modern theory of transformational linguistics and the information-processing theory of cognition were born in the same matrix—the matrix of ideas produced by the development of the modern digital computer, and in the realization that, though the computer was embodied in hardware, its soul was a program. One of the initial professional papers on transformational linguistics and one of the initial professional papers on information-processing psychology were presented, the one after the other, at a meeting at M.I.T. in September 1956.[18] Thus the two bodies of theory have had cordial relations from an early date. And quite rightly, for they rest conceptually on the same view of the human mind.

Now some may object that this is not correct and that they rest on almost diametrically opposed views of the human mind. For I have stressed the artificial character of human thinking—how it adapts itself, through individual learning and social transmission of knowledge, to the requirements of the task environment. The leading exponents of the formal linguistic theories, on the other hand, have taken what is sometimes called a "nativist" position. They have argued that a child could never acquire any skill so complex as speaking and understanding language if he did not already have built into him at birth the basic machinery for the exercise of these skills.

The issue is reminiscent of the debate on language universals—on whether there are some common characteristics shared by all known tongues. We know that the communalities among languages are not in any sense specific but that they relate, instead, to very broad

[18] N. Chomsky, "Three Models for the Description of Language," and A. Newell and H. A. Simon, "The Logic Theory Machine," both in *IRE Transactions on Information Theory*, IT-2, No. 3 (September 1956).

structural characteristics that all languages seem to share in some manner. Something like the distinction between noun and verb—between object and action or relation—appears to be present in all human languages. All languages appear to have the boxes-within-boxes character called phrase structure, All languages appear to derive certain strings from others by transformation.[19]

Now if we accept these as typical of the universals to which the nativist argument appeals, there are still at least two different possible interpretations of that argument. The one is that the language competence is *purely* linguistic, that language is *sui generis*, and that the human faculties it calls upon are not all employed also in other performances.

An alternative interpretation of the nativist position is that producing utterances and understanding the utterances of others depend on some characteristics of the human central nervous system which are common in all languages but also essential to other aspects of human thinking besides speech and listening.

The former interpretation does not, but the latter does, provide an explanation for the remarkable parallelism holding between the underlying assumptions about human capabilities that are embedded in modern linguistic theory and the assumptions embedded in information-processing theories of human thinking. The kinds of assumptions that I made earlier about the structure of human memory are just the kinds of assumptions one would want to make for a processing system capable of handling language. Indeed, there has been extensive borrowing back and forth between the two fields. Both postulate hierarchically organized list structures as a basic principle of memory organization. Both are concerned with how a serially operating processor can convert strings of symbols into list structures or list structures into strings. In both fields, the same general classes of computer-programming languages have proved convenient for modeling and simulating the phenomena.

[19] On language universals see Joseph H. Greenberg (ed.), *Universals of Language* (Cambridge: M.I.T. Press, 1963), particularly Greenberg's own chapter, pp. 58–90. On the "nativist" position, see Jerrold J. Katz, *The Philosophy of Language* (New York: Harper & Row, 1966), pp. 240–282.

Semantics in Language Processing

Let me suggest one way in which the relation between linguistic theories and information-processing theories of thinking is going to be even closer in the future than it was in the past. Linguistic theory has thus far been largely a theory of syntax, of grammar. In practical application to such tasks as automatic translation, it has encountered difficulties when translation depended on more than syntactic cues—when it depended on context and "meanings." It seems pretty clear that one of the major directions that progress in linguistics will have to take is toward development of an adequate semantics to complement syntax.

The theory of thinking I have been outlining can already provide an important part of such a semantic component. The principles of memory organization I have described can be used as a basis both for discussing the internal representation of linguistic strings and for discussing the internal representation of two-dimensional visual stimuli, or other nonlinguistic stimuli. Given these comparable bases for the organization of the several kinds of stimuli, it becomes easier to conceptualize the cooperation of syntactic and semantic cues in the interpretation of language.

Several research projects have been carried out at M.I.T., and several at Carnegie-Mellon University, in recent years that bear on this point. I should like to mention just two of these, which illustrate how this approach might be used to explain the resolution of syntactic ambiguities by use of semantic cues.

L. Stephen Coles, in a dissertation completed in 1967,[20] described a computer program that uses pictures on a cathode ray tube to resolve syntactic ambiguities. I shall paraphrase his procedure with an example that is easier to visualize than any he actually used. Consider the sentence:

I saw the man on the hill with the telescope.

[20] L. Stephen Coles, *Syntax Directed Interpretation of Natural Language*, doctoral dissertation, Carnegie Institute of Technology, 1967.

This sentence has at least three acceptable interpretations; a linguist could, no doubt, discover others. Which of the three obvious ones we pick depends on where we think the telescope is: Do I have it? Does the man on the hill have it? Or is it simply on the hill, not in his hands?

Now suppose that the sentence is accompanied by Figure 5. The issue is no longer in doubt. Clearly, it is I who have the telescope.

Coles's program is capable of recognizing objects in a picture, and relations among objects; and it is capable of representing the picture as a list structure, which, in the example before us, we might describe thus:

SAW ((I, WITH (telescope)), (man, ON (hill)))

I have not tried to reproduce the actual details of the scheme he used, but I have simply shown that a picture, so represented, could readily be matched against alternate parsings of a verbal string and thus used to resolve the ambiguity of the latter.

Another program, recently completed by Laurent Siklossy (personal communication), illustrates how semantic information can aid in the *acquisition* of a language. The reader may be familiar with the "Language Through Pictures" books developed by Professor I. A. Richards and his associates. These books have been prepared for a large number of languages. On each page is a picture and beneath it one or more sentences that say something about the picture in the language to be learned. The sequence of pictures and accompanying sentences is arranged to proceed from very simple situations ("I am here," "That is a man") to more complex ones ("The book is on the shelf").

Siklossy's program takes as its input an analogue to one of the "Language Through Pictures" books. The picture is assumed to have already been transformed into a list structure (not unlike the one illustrated earlier for Coles's system) as its internal representation. The program's task is to learn, when confronted with such a picture, to utter the appropriate sentence in the natural language it is learning—a sentence that says what the picture shows. In the case of the sentence about the telescope (somewhat more complicated than any on

Figure 5. A Syntactically Ambiguous Sentence: "I saw the man on the hill with the telescope"

which the scheme has actually been tested), one would hope that the program would respond to the picture with "I saw the man on the hill with the telescope," if it were learning English, or *Ich habe den Mann auf dem Berg mit dem Fernglas gesehen*, if it were learning German.

Of course, the program could respond correctly only if it already knew the vocabulary and grammar required for the translation. In other cases, it would use the sentence associated with the picture to add to its vocabulary and syntax.

I do not wish to expand some pioneering experiments into a comprehensive theory of semantics. The point of these examples is that they show that the kind of memory structure that has been postulated, for other reasons, to explain human behavior in simpler cognitive tasks is suitable for explaining how linguistic strings might be represented internally, how other kinds of stimuli might be similarly represented, and how the communalities in representation—the use of hierarchically organized list structures for both—may explain how language and "meanings" come together in the human head.

There is no contradiction, then, between the thesis that a human being possesses, at birth, a competence for acquiring and using language and the thesis that language is the most artificial, hence also the most human of all human constructions. The former thesis is an assertion that there *is* an inner environment and that it does place limits on the kinds of information processing of which the organism is capable. The structure of language reveals these limits; and these limits, in turn, account for such commonality as exists among the Babel of human tongues.

The latter thesis, of the artificiality of language, is an assertion that the limits on adaptation, on possible languages, imposed by the inner environment are very broad limits on organization, not very specific limits on syntax. Moreover, according to the thesis, they are limits imposed not only on language but also on every other mode of representing internally experience received through stimuli from outside.

Such a view of the relation of language and thinking puts a new cast on the "Whorfian" hypothesis that—stating it in overstrong form—only the expressible is thinkable. If the view is valid, it would be as correct to say, "Only the thinkable is expressible"—a view that, I suppose, Kant would have found quite congenial.

Conclusion

The thesis with which I began this talk was the following:

A man, viewed as a behaving system, is quite simple. The apparent complexity of his behavior over time is largely a reflection of the complexity of the environment in which he finds himself.

That hypothesis was based, in turn, on the thesis of the previous chapter: that behavior is adapted to goals, hence is artificial, hence reveals only those characteristics of the behaving system that limit the adaptation.

To illustrate how we have begun to test these theses and at the same time build up a theory of the simple principles that underlie human behavior, I have surveyed some of the evidence from a range of human performances, particularly those that have been studied in the psychological laboratory.

The behavior of human subjects in solving cryptarithmetic problems, in attaining concepts, in memorizing, in holding information in short-term memory, in processing visual stimuli, and in performing tasks that use natural languages provides strong support for these theses. The artificiality—hence variability—of human behavior hardly calls for evidence beyond our observation of everyday life. The experiments are therefore mostly significant in what they show about the broad communalities in organization of the human information-processing system as it engages in different tasks.

The evidence is overwhelming that the system is basically serial in its operation: that it can process only a few symbols at a time and that the symbols being processed must be held in special, limited memory structures whose content can be changed rapidly. The most striking limits on subjects' capacities to employ efficient strategies arise from the very small capacity of the short-term memory structure (seven chunks) and from the relatively long time (five seconds) required to transfer a chunk of information from short-term to long-term memory.

When we turn from tasks that exercise mainly the short-term memory and serial-processing capabilities of the central nervous system to tasks that involve retrieval of stored information, we encounter new limits of adaptation, and through these limits we acquire new information about the organization of mind and brain. Studies of visual perception and of tasks requiring use of natural language show with growing clarity that memory is indeed organized in associative fashion, but that the "associations" have the properties of what, in the computer trade, are usually called "list structures." I have indicated briefly what those properties are.

These are the sorts of generalizations about human thinking that are emerging from the experimental evidence. They are simple things, just as our hypothesis led us to expect. Moreover, though the picture will continue to be enlarged and clarified, we should not expect it to become essentially more complex. Only human pride argues that the apparent intricacies of our path stem from a quite different source than the intricacy of the ant's path.

One of the curious consequences of my approach—of my thesis—is that I have said nothing about physiology. But the mind is usually thought to be located in the brain. I have discussed the organization of the mind without saying anything about the structure of the brain.

The main reason for this disembodiment of mind is, of course, the thesis that I have just been discussing. The difference between the hardware of a computer and the "hardware" of the brain has not prevented computers from simulating a wide spectrum of kinds of human thinking—just because both computer and brain, when engaged in thought, are adaptive systems, seeking to mold themselves to the shape of the task environment.

It would be unfortunate if this conclusion were altered to read that neurophysiology has nothing to contribute to the explanation of human behavior. That would be, of course, a ridiculous doctrine. But our analysis of the artificial leads us to a particular view of the form that the physiological explanation of behavior must take. Neurophysiology is the study of the inner environment of the adaptive system called Homo sapiens. It is to physiology that we must turn for an explanation of the limits of adaptation: Why is short-term memory limited to seven chunks; what is the physiological structure that corresponds to a "chunk"; what goes on during the five seconds that a chunk is being fixated; how are associational structures realized in the brain?

As our knowledge increases, the relation between physiological and information-processing explanations will become just like the relation between quantum-mechanical and physiological explanations in biology (or the relation between solid-state physics and programming explanations in computer science). They constitute two linked levels of explanation with (in the case before us) the limiting properties of the inner system showing up at the interface between them.

Finally, we may expect also that as we link information-processing psychology to physiology on the inner side, we shall also be linking psychology to the general theory of search through large combinatorial spaces on the outer side—the side of the task environment. But that brings me to the topic of my third chapter, for the theory of design *is* that general theory of search.

3 THE SCIENCE OF DESIGN: Creating the Artificial

HISTORICALLY and traditionally, it has been the task of the science disciplines to teach about natural things: how they are and how they work. It has been the task of engineering schools to teach about artificial things: how to make artifacts that have desired properties and how to design.

Engineers are not the only professional designers. Everyone designs who devises courses of action aimed at changing existing situations into preferred ones. The intellectual activity that produces material artifacts is no different fundamentally from the one that prescribes remedies for a sick patient or the one that devises a new sales plan for a company or a social welfare policy for a state. Design, so construed, is the core of all professional training; it is the principal mark that distinguishes the professions from the sciences. Schools of engineering, as well as schools of architecture, business, education, law, and medicine, are all centrally concerned with the process of design.

In view of the key role of design in professional activity, it is ironic that in this century the natural sciences have almost driven the sciences of the artificial from professional school curricula. Engineering schools have become schools of physics and mathematics; medical schools have become schools of biological science; business schools have become schools of finite mathematics. The use of adjectives like "applied" conceals, but does not change, the fact. It simply means that in the professional schools those topics are selected from mathematics and the natural sciences for emphasis which are thought to be most nearly relevant to professional practice. It does not mean that design is taught, as distinguished from analysis.

The movement toward natural science and away from the sciences of the artificial has proceeded further and faster in engineering, business, and medicine than in the other professional fields I have mentioned, though it has by no means been absent from schools of law, journalism, and library science. The stronger universities are more deeply affected than the weaker, and the graduate programs more than the undergraduate. Few doctoral dissertations in first-rate professional schools today deal with genuine design problems, as distinguished from problems in solid-state physics or stochastic processes. I have to make partial exceptions—for reasons I shall mention—of dissertations in computer science and management science, and there are undoubtedly some others, for example, in chemical engineering.

Such a universal phenomenon must have a basic cause. It does have a very obvious one. As professional schools, including the independent engineering schools, are more and more absorbed into the general culture of the university, they hanker after academic respectability. In terms of the prevailing norms, academic respectability calls for subject matter that is intellectually tough, analytic, formalizable, and teachable. In the past, much, if not most, of what we knew about design and about the artificial sciences was intellectually soft, intuitive, informal, and cookbooky. Why would anyone in a university stoop to teach or learn about designing machines or planning market strategies when he could concern himself with solid-state physics? The answer has been clear: he usually wouldn't.

The problem is widely recognized in engineering and medicine, today, and to a lesser extent in business. Some do not think it a problem, because they regard schools of applied science as a superior alternative to the trade schools of the past. If that were the choice, we could agree.[1] But neither alternative is satisfactory. The older kind of professional school did not know how to educate for professional design at an intellectual level appropriate to a university; the newer kind of school has nearly abdicated responsibility for training in the core professional skill. Thus we are faced with a problem of devising a professional school that can attain two objectives simultaneously: education in both artificial and natural science at a high intellectual level. This too is a problem of design—organizational design.

The kernel of the problem lies in the phrase "artificial science." In my two previous chapters I have shown that a science of artificial phenomena is always in imminent danger of dissolving and vanishing. The peculiar properties of the artifact lie on the thin interface between the natural laws within it and the natural laws without. What can we say about it? What is there to study besides the boundary

[1] That was, in fact, the choice in our engineering schools a generation ago. The schools needed to be purged of vocationalism; and a genuine science of design did not exist, even in a rudimentary form, as an alternative. Hence, the road forward was the road toward introducing more fundamental science. Karl Taylor Compton was one of the prominent leaders in this reform, which was a main theme in his presidential inaugural address at M.I.T. in 1930: "I hope . . . that increasing attention in the Institute may be given to the fundamental sciences; that they may achieve as never before the spirit and results of research; that all courses of instruction may be examined carefully to see where training in details has been unduly emphasized at the expense of the more powerful training in all-embracing fundamental principles."

Notice that President Compton's emphasis was on "fundamental," an emphasis as sound today as it was in 1930. What I am urging in this essay is not a departure from the fundamental but an inclusion in the curriculum of the fundamental in engineering along with the fundamental in natural science. That was not possible in 1930; but it is possible today.

sciences—those that govern the means and the task environment?

The artificial world is centered precisely on this interface between the inner and outer environments; it is concerned with attaining goals by adapting the former to the latter. The proper study of those who are concerned with the artificial is the way in which that adaptation of means to environments is brought about—and central to that is the process of design itself. The professional schools will reassume their professional responsibilities just to the degree that they can discover a science of design, a body of intellectually tough, analytic, partly formalizable, partly empirical, teachable doctrine about the design process.

It is the thesis of this chapter that such a science of design not only is possible but is actually emerging at the present time. It has already begun to penetrate the engineering schools, particularly through programs in computer science and "systems engineering," and business schools through management science. Perhaps it also has beachheads in other professional curricula, but these are the two with which I am most familiar. We can already see enough of its shape to predict some of the important ways in which engineering schools tomorrow will differ from departments of physics, and business schools from departments of economics and psychology. Let me now turn from questions of university organization to the substance of the matter.

The Logic of Design: Fixed Alternatives

We must start with some questions of logic.[2] The natural sciences are concerned with how things are. Ordinary systems of logic—the standard propositional and predicate calculi, say—serve these sciences well. Since the concern of standard logic is with declarative statements, it is well

[2] I have treated the question of logical formalism for design at greater length in two earlier papers: "The Logic of Rational Decision," *British Journal for the Philosophy of Science*, *16*: 169–186 (1965); and "The Logic of Heuristic Decision Making," in Nicholas Rescher (ed.), *The Logic of Decision and Action* (Pittsburgh: University of Pittsburgh Press, 1967), pp. 1–35. The present discussion is based on these two papers.

suited for assertions about the world and for inferences from those assertions.

Design, on the other hand, is concerned with how things ought to be, with devising artifacts to attain goals. We might question whether the forms of reasoning that are appropriate to natural science are suitable also for design. One might well suppose that introduction of the verb "should" may require additional rules of inference, or modification of the rules already imbedded in declarative logic.

Paradoxes of Imperative Logic

Various "paradoxes" have been constructed to demonstrate the need for a distinct logic of imperatives, or a normative, deontic logic. In ordinary logic, from "Dogs are pets" and "Cats are pets," one can infer "Dogs and cats are pets." But from "Dogs are pets," "Cats are pets," and "You should keep pets," can one infer "You should keep cats and dogs"? And from "Give me needle and thread!" can one deduce, in analogy with declarative logic, "Give me needle or thread!"? Easily frustrated people would perhaps rather have neither needle nor thread than one without the other; and peace-loving people, neither cats nor dogs, rather than both.

As a response to these challenges of apparent paradox, there have been developed a number of constructions of modal logic for handling "shoulds," "shalts," and "oughts" of various kinds. I think it is fair to say that none of these systems has been sufficiently developed or sufficiently widely applied to demonstrate that it is adequate to handle the logical requirements of the process of design.

Fortunately, such a demonstration is really not essential, for it can be shown that the requirements of design can be met fully by a modest adaptation of ordinary declarative logic. Thus, a special logic of imperatives is unnecessary.[3]

[3] I should like to underline the word "unnecessary." When I said something like this in another place (the second paper mentioned in the previous footnote), an able logician, who had specialized in modal logics, accused me of asserting that modal logics were "impossible." Now this is patently false: modal logics can be shown to exist in the same way that giraffes can—namely, by exhibiting some of them. The question is not whether they exist but what they are good for. A modal logician should have no difficulty in distinguishing "non-necessity" from "impossibility."

Reduction to Declarative Logic

The easiest way to discover what kinds of logic are needed for design is to examine what kinds of logic designers use when they are being careful about their reasoning. Now there would be no point in doing this if designers were always sloppy fellows who reasoned loosely, vaguely, and intuitively. Then we might say that whatever logic they used was not the logic they *should* use.

However, there exists a considerable area of design practice where standards of rigor in inference are as high as one could wish. I refer to the domain of so-called "optimization methods," most highly developed in statistical decision theory and management science but acquiring growing importance also in engineering design theory. The theories of probability and utility, and their intersection, have received the painstaking attention not only of practical designers and decision makers but also of a considerable number of the most distinguished logicians and mathematicians of the present and recent past generations. F. P. Ramsey, B. de Finetti, A. Wald, J. von Neumann, J. Neyman, K. Arrow, and L. J. Savage are examples.

The logic of optimization methods can be sketched as follows: The "inner environment" of the design problem is represented by a set of given alternatives of action. The alternatives may be given *in extenso:* more commonly, they are specified in terms of *command variables* that have defined domains. The "outer environment" is represented by a set of parameters, which may be known with certainty or only in terms of a probability distribution. The goals for adaptation of inner to outer environment are defined by a utility function—a function, usually scalar, of the command variables and environmental parameters—perhaps supplemented by a number of constraints (inequalities, say, between functions of the command variables and environmental parameters). The optimization problem is to find an admissible set of values of the command variables, compatible with the constraints, that maximize the utility function for the given values of the environmental parameters. (In the probabilistic case, we might say, "maximize the expected value of the utility function," for instance, instead of "maximize the utility function.")

Logical Terms		*Example: The Diet Problem*
command variables	("means")	quantities of foods
fixed parameters	("laws")	prices of foods; nutritional contents
constraints	("ends")	nutritional requirements
utility function	("ends")	−cost of diet

Constraints characterize the inner environment;
parameters characterize the outer environment.

Problem: Given the constraints and fixed parameters; find values of the command variables that maximize utility.

Figure 6. The Paradigm for Imperative Logic

A stock application of this paradigm is the so-called "diet problem," shown in Figure 6. A list of foods is provided, the command variables being quantities of the various foods that are to be included in the diet. The environmental parameters are the prices and nutritional contents (calories, vitamins, minerals, and so on) of each of the foods. The utility function is the cost (with a minus sign attached) of the diet, subject to the constraints, say, that it not contain more than 2,000 calories per day, that it meet specified minimum needs for vitamins and minerals, and that rutabaga not be eaten more than once a week. The constraints may be viewed as characterizing the inner environment. The problem is to select the quantities of foods that will meet the nutritional requirements and side conditions at the given prices for the lowest cost.

The diet problem is a simple example of a class of problems that are readily handled, even when the number of variables is exceedingly large, by the mathematical formalism known as linear programming. I shall come back to the technique a little later. My present concern is with the logic of the matter.

Since the optimization problem, once formalized, is a standard mathematical problem—to maximize a function subject to constraints—it is evident that the logic used to deduce the answer is the standard logic of the predicate calculus on which mathematics rests. How does the formalism avoid making use of a special logic of imperatives? It does so by dealing with sets of *possible worlds:* First consider all the possible worlds that meet the constraints of the outer environment; then find the particular world in the set that meets the remaining constraints of the goal and maximizes the utility function. The logic is exactly the same as if we were to adjoin the goal constraints and the maximization requirement, as new "natural laws," to the existing natural laws embodied in the environmental conditions.[4] We simply ask what values the command variables *would* have in a world meeting all these conditions and conclude that these are the values the command variables *should* have.

Computing the Optimum

Our discussion thus far has already provided us with two central topics for the curriculum in the science of design:

1. *Utility theory and statistical decision theory—a logical framework for rational choice among given alternatives.*
2. *The body of techniques for actually deducing which of the available alternatives is the optimum.*

Only in trivial cases is the computation of the optimum alternative an easy matter. If utility theory is to have application to real-life design problems, it must be accompanied by tools for actually making the computations. The dilemma of the rational chess player is familiar to all.

[4] The use of the notion of "possible worlds" to embed the logic of imperatives in declarative logic goes back at least to Jørgen Jørgensen, "Imperatives and Logic," *Erkenntnis*, *7*: 288–296 (1937–1938). See also my *Administrative Behavior* (New York: Macmillan, 1947), Chapter III. More recently, this same idea has been used by several logicians to construct a formal bridge between the predicate calculus and modal logic by means of so-called semantic or model-theoretic methods. See, for example, Richard Montague, "Logical Necessity, Physical Necessity, Ethics, and Quantifiers," *Inquiry*, *4*: 259–269 (1960), where references are also given to work of Stig Kanger and Saul Kripke; and Jaakko Hintikka, "Modality and Quantification," *Theoria*, *27*: 119–128 (1961). While these model-theoretic proposals are basically sound, none of them seems yet to have given adequate attention to the special role played in the theory by command variables and criterial constraints.

The optimal strategy in chess is easily demonstrated: simply assign a value of + 1 to a win, 0 to a draw, −1 to a loss; consider all possible courses of play; minimax backward from the outcome of each, assuming each player will take the most favorable move open to him at any given point. This procedure will determine what move to make now. The only trouble is that the computations required are astronomical (the number 10^{120} is often mentioned in this context) and hence cannot be carried out—not by men, not by existing computers, not by prospective computers.

A theory of design as applied to the game of chess would encompass not only the utopian minimax principle but also some practicable procedures for finding good moves in actual board positions in real time, within the computational capacities of real men or real computers. No exceptionally good procedures of this kind exist today, other than those stored in the memories of grandmasters, but there is at least one computer program, due principally to Greenblatt,[5] that plays a respectable game in week-end tournaments and has a Class C American Chess Federation rating. Analogous procedures do exist in other task areas that have perhaps more practical importance than chess.

The second topic, then, for the curriculum in the science of design consists in the efficient computational techniques that are available for actually finding optimum courses of action in real situations, or reasonable approximations to real situations. That topic has a number of important components today, most of them developed—at least to the level of practical application—within the past 25 years. These include linear programming theory, dynamic programming, geometric programming, queuing theory, and control theory.

Finding Satisfactory Actions

The subject of computational techniques need not be limited to optimization. Traditional engineering design

[5] Richard D. Greenblatt, with Donald E. Eastlake III and Stephen D. Crocker, "The Greenblatt Chess Program," *Proceedings of the Fall Joint Computer Conference, 1967* (Anaheim, California), pp. 801–810.

methods make much more use of inequalities—specifications of satisfactory performance—than of maxima and minima. So-called "figures of merit" permit comparison between designs in terms of "better" and "worse" but seldom provide a judgment of "best." For example, I may cite the root-locus methods employed in the design of servomechanisms.

Since there did not seem to be any word in English for decision methods that look for good or satisfactory solutions instead of optimal ones, some years ago I introduced the term "satisficing" to refer to such procedures. Now, no one in his right mind will satisfice if he can equally well optimize; no one will settle for good or better if he can have best. But that is not the way the problem usually poses itself in actual design situations.

In the real world we usually do not have a choice between satisfactory and optimal solutions, for we only rarely have a method of finding the optimum. Consider, for example, the well-known combinatorial problem called the traveling saleman problem: Given the geographical locations of a set of cities, find the routing that will take a salesman to all the cities with the shortest mileage. For this problem there is a straightforward optimizing algorithm (analogous to the minimax algorithm for chess): try all possible routings, and pick the shortest. But for any considerable number of cities, the algorithm is computationally infeasible (the number of routes through N cities will be $N!$). Although some ways have been found for cutting down the length of the search, no algorithm has been discovered sufficiently powerful to solve the traveling saleman problem with a tolerable amount of computing for a set of, say, fifty cities.

Rather than keep our salesman at home, we shall prefer, of course, to find a satisfactory, if not optimal, routing for him. Under most circumstances, common sense will probably arrive at a fairly good route, but an even better one can often be found by one or another of several heuristic methods.[6]

[6] "The traveling salesman problem" and a number of closely analogous combinatorial problems—such as the "warehouse location problem"—have considerable practical importance—for instance, in siting central power stations for an interconnected grid.

An earmark of all these situations where we satisfice for inability to optimize is that, although the set of available alternatives is "given" in a certain abstract sense (we can define a generator guaranteed to generate all of them eventually), it is not "given" in the only sense that is practically relevant. We cannot, within practicable computational limits, generate all the admissible alternatives and compare their respective merits. Nor can we recognize the best alternative, even if we are fortunate enough to generate it early, until we have seen all of them. We satisfice by looking for alternatives in such a way that we can generally find an acceptable one after only moderate search.

Now in many satisficing situations, the expected length of search for an alternative meeting specified standards of acceptability depends on how high the standards are set, but it depends hardly at all on the total size of the universe to be searched. The time required for a search through a haystack for a needle sharp enough to sew with depends on the density of distribution of sharp needles, but not on the total size of the stack.

Hence, when we use satisficing methods, it often does not matter whether or not the total set of admissible alternatives is "given" by a formal but impracticable algorithm. It often does not even matter how big that set is. For this reason, satisficing methods may be extendable to design problems in that broad range where the set of alternatives is not "given" even in the Quixotic sense that it is "given" for the traveling salesman problem. Our next task is to examine this possibility.

The Logic of Design: Finding Alternatives

When we take up the case where the design alternatives are not given in any constructive sense but must be synthesized, we must ask once more whether any new forms of reasoning are involved in the synthesis, or whether, again, the standard logic of declarative statements is all we need.

In the case of optimization, we asked: "Of all possible worlds (those attainable for some admissible values of the action variables), which is the best (yields the highest value of the criterion function)?" As we saw, this is a purely empirical question, calling only for facts and ordinary declarative reasoning to answer it.

In this case, where we are seeking a satisfactory alternative, once we have found a candidate we can ask: Does this alternative satisfy all the design criteria? Clearly, this is also a factual question and raises no new issues of logic. But how about the process of *searching* for candidates? What kind of logic is needed for the search?

Means-End Analysis

The condition of any goal-seeking system is that it is connected to the outside environment through two kinds of channels: the afferent, or sensory, channels, through which it receives information about the environment; and the efferent, or motor, channels, through which it acts on the environment.[7] The system must have some means of storing in its memory information about states of the world—afferent, or sensory, information—and information about actions—efferent, or motor, information. Ability to attain goals depends on building up associations, which may be simple or very complex, between particular changes in states of the world and particular actions that will (reliably or not) bring these changes about.

Except for a few built-in reflexes, an infant has no basis for correlating his sensory information with his actions. A very important part of his early learning is that particular actions or sequences of actions will bring about particular changes in the state of the world as he senses it. Until he builds up this knowledge, the world of sense and the motor world are two entirely separate, entirely unrelated worlds. Only as he begins to acquire experience as to how elements of the one relate to elements of the other can he act purposefully on the world.

The computer problem-solving program called GPS, designed to model some of the main features of human

[7] Notice that we are not saying that the two kinds of channels operate independently of each other, since they surely do not in living organisms, but that we can distinguish conceptually, and to some extent neurologically, between the incoming and outgoing flows.

problem solving, exhibits in stark form how goal-directed action depends on building this kind of bridge between the afferent and the efferent worlds. On the afferent, or sensory, side, GPS must be able to represent desired situations or desired objects, as well as the present situation. It must be able, also, to represent *differences* between the desired and the present. On the efferent side, GPS must be able to represent *actions* that change objects or situations. In order to behave purposefully, GPS must be able to select, from time to time, those particular actions that are likely to reduce or remove the particular differences between desired and present that the system detects. In the machinery of GPS, this selection is achieved through a *table of connections*, which associates with each kind of detectable difference the actions that are relevant to reducing that difference. These are its associations, which relate the afferent to the efferent world. Since reaching a goal generally requires a sequence of actions, and since some attempts may be ineffective, GPS must also have means for detecting the progress it is making (the changes in the differences between the actual and the desired) and for trying alternate paths.

The Logic of Search

GPS, then, is a system that searches selectively through a (possibly large) environment in order to discover and assemble sequences of actions that will lead it from a given situation to a desired situation. What are the rules of logic that govern such a search? Is anything more than standard logic involved? Do we require a modal logic to rationalize the process?

Standard logic would seem to suffice. To represent the relation between the afferent and the efferent worlds, we conceive GPS as moving through a large maze. The nodes of the maze represent situations, described afferently; the paths joining one node to another are the actions, described as motor sequences, that will transform the one situation into the other. At any given moment, GPS is always faced with a single question: "What action shall I try next?" Since GPS has some imperfect knowledge about the relations of actions to changes in the situation, this becomes a question of choice under uncertainty of a kind already discussed in a previous section.

It is characteristic of the search for alternatives that the solution, the complete action that constitutes the final design, is built from a sequence of component actions. The enormous size of the space of alternatives arises out of the innumerable ways in which the component actions, which need not be very numerous, can be combined into sequences.

Much is gained by considering the component actions in place of the sequences that constitute complete actions, because the situation when viewed afferently usually factors into components that match at least approximately the component actions derived from an efferent factorization. The reasoning implicit in GPS is that if a desired situation differs from a present situation by differences $D_1, D_2, \ldots, D_n$, and if action A_1 removes differences of type D_1, action A_2 removes differences of type D_2, and so on, then the present situation can be transformed into the desired situation by performing the sequence of actions $A_1A_2 \ldots A_n$.

This reasoning is by no means valid, in terms of the rules of standard logic, in all possible worlds. Its validity requires some rather strong assumptions about the independence of the effects of the several actions on the several differences. One might say that the reasoning is valid in worlds that are "additive" or "factorable" in a certain sense. (The air of paradox about the cat-dog and needle-thread examples cited earlier arises precisely from the nonadditivity of the actions in these two cases. The first is, in economists' language, a case of decreasing returns, the second a case of increasing returns.)

Now the real worlds to which problem solvers and designers address themselves are seldom completely additive in this sense. Actions have side consequences (may create new differences) and sometimes can only be taken when certain side conditions are satisfied (call for removal of other differences before they become applicable). Under these circumstances, one can never be certain that a partial sequence of actions that accomplishes *certain* goals can be augmented to provide a solution that satisfies *all* the conditions and attains *all* the goals (even though they be satisficing goals) of the problem.

For this reason, problem-solving systems and design procedures in the real world do not merely *assemble* problem solutions from components but must *search* for appropriate assemblies. In carrying out such a search, it is often efficient to divide one's eggs among a number of baskets—that is, not to follow out one line until it succeeds completely or fails definitely, but to begin to explore several tentative paths, continuing to pursue a few that look most promising at a given moment. If one of the active paths begins to look less promising, it may be replaced by another that had previously been assigned a lower priority.

Our discussion of design when the alternatives are not given has yielded at least three additional topics for instruction in the science of design:

3. *Adaptation of standard logic to the search for alternatives.* Design solutions are sequences of actions that lead to possible worlds satisfying specified constraints. With satisficing goals, the sought-for possible worlds are seldom unique; the search is for *sufficient*, not *necessary*, actions for attaining goals.

4. *The exploitation of parallel, or near-parallel, factorizations of differences.* Means-end analysis is an example of a broadly applicable problem-solving technique that exploits this factorization.

5. *The allocation of resources for search to alternative partly explored action sequences.* I should like to elaborate somewhat on this last-mentioned topic.

Design as Resource Allocation

There are two ways in which design processes are concerned with the allocation of resources. First, conservation of scarce resources may be one of the criteria for a satisfactory design. Second, the design process itself involves management of the resources of the designer, so that his efforts will not be dissipated unnecessarily in following lines of inquiry that prove fruitless.

There is nothing special that needs to be said here about resource conservation—cost minimization, for example, as a design criterion. Cost minimization has always been an implicit consideration in the design of engineering structures, but until a few years ago it generally *was* only implicit, rather than explicit. More and more, cost calculations have been brought explicitly into the design procedure, and a strong case can be made today for training design engineers in that body of technique and theory that economists know as "cost-benefit analysis."

An Example from Highway Design

The notion that the costs of designing must themselves be considered in guiding the design process is one that has been introduced much more recently and that has still not had wide application, except at an intuitive level. A good example of what I mean by this is the procedure, developed by Marvin L. Manheim as a doctoral thesis at M.I.T., for solving highway location problems.[8]

Manheim's procedure incorporates two main notions: first, the idea of specifying a design progressively from the level of very general plans down to determining the actual construction; second, the idea of attaching values to plans at the higher levels as a basis for deciding which plans to pursue to levels of greater specificity.

In the case of highway design, the higher-level search is directed toward discovering "bands of interest" within which the prospects of finding a good specific route are promising. Within each band of interest, one or more locations is selected for closer examination. Specific designs are then developed for particular locations. The scheme is not limited, of course, to this specific three-level division, but it can be generalized as appropriate.

Manheim's scheme for deciding which alternatives to pursue from one level to the next is based on assigning costs to each of the design activities and estimating highway costs for each of the higher-level plans. The highway cost associated with a plan is a prediction of what the cost would be for the actual route if that plan were particularized through subsequent design activity. In other words, it is a measure of how "promising" a

[8] Marvin L. Manheim, *Hierarchical Structure: A Model of Design and Planning Processes* (Cambridge: The M.I.T. Press, 1966).

plan is. Those plans are then pursued to completion that look most promising after the prospective design costs have been offset against them.

In the particular method that Manheim describes, the "promise" of a plan is represented by a probability distribution of outcomes that would ensue if it were pursued to completion. The distribution must be estimated by the engineer—a serious weakness of the method—but, once estimated, it can be used within the framework of Bayesian decision theory. The particular probability model used is not the important thing about the method; other methods of valuation, without the Bayesian superstructure, might be just as satisfactory.

In the highway location procedure, the evaluation of higher-level plans performs two functions. First, it answers the question "Where shall I search next?" Second, it answers the question "When shall I stop the search and accept a solution as satisfactory?" Thus it is both a steering mechanism for the search and a satisficing criterion for terminating the search.

Schemes for Guiding Search

Let us generalize the notion of schemes for guiding search activity beyond Manheim's specific application to a highway location problem and beyond his specific guidance scheme based on Bayesian decision theory. Consider the typical structure of a problem-solving program. The program begins to search along possible paths, storing in memory a "tree" of the paths it has explored. Attached to the end of each branch—each partial path—is a number that is supposed to express the "value" of that path.

But the term "value" is really a misnomer. A partial path is not a solution of the problem, and a path has a "true" value of zero unless it leads toward a solution. Hence it is more useful to think of the values as estimates of the gain to be expected from further search along the path than to think of them as "values" in any more direct sense. For example, it may be desirable to attach a relatively high value to a partial exploration that *may* lead to a very good solution but with a low probability.

If the prospect fades on further exploration, only the cost of the search has been lost. The disappointing outcome need not be accepted, but an alternative path may be taken instead. Thus the scheme for attaching values to partial paths may be quite different from the evaluation function for proposed complete solutions.[9]

When we recognize that the purpose of assigning values to incomplete paths is to guide the choice of the next point for exploration, it is natural to generalize even further. All kinds of information gathered in the course of search may be of value in selecting the next step in search. We need not limit ourselves to valuations of partial search paths.

For example, in a chess-playing program, an exploration may generate a continuation move different from any that was proposed by the initial move generator. Whatever the context—the branch of the search tree—on which the move was actually generated, it can now be removed from that context and considered in the context of other move sequences. Such a scheme, on a limited basis, was added by Baylor to MATER, a program for discovering checkmating combinations in chess; and it proved to enhance the program's power significantly.[10]

Thus, search processes may be viewed—as they have been in most discussions of problem solving—as processes for seeking a problem solution. But they can be viewed more generally as processes for gathering information about problem structure that will ultimately be valuable in discovering a problem solution. The latter viewpoint is more general than the former in a significant sense, in that it suggests that information obtained along any particular branch of a search tree may be used in many contexts besides the one in which it was generated. Only a few problem-solving programs exist today that can be regarded as moving even a modest distance from the earlier, more

[9] That this point is not obvious can be seen from the fact that most chess-playing programs have used similar or identical evaluation procedures both to guide search and to evaluate the positions reached at the ends of paths.

[10] George W. Baylor and Herbert A. Simon, "A Chess Mating Combinations Program," *Proceedings of the Spring Joint Computer Conference, 1966, 28*: 431–447 (Boston, April 26–28) (Washington: Spartan Books, 1966).

limited viewpoint to the newer one. Here is an important direction for research in the theory of design.

THE SHAPE OF THE DESIGN: HIERARCHY

In my first chapter, I gave some reasons why complex systems might be expected to be constructed in a hierarchy of levels, or in a boxes-within-boxes form. The basic idea is that the several components in any complex system will perform particular subfunctions that contribute to the over-all function. Just as the "inner environment" of the whole system may be defined by describing its functions, and without detailed specification of its mechanisms, so the "inner environment" of each of the subsystems may be defined by describing the functions of that subsystem, and without detailed specification of *its* submechanisms.[11]

To design such a complex structure, one powerful technique is to discover viable ways of decomposing it into semi-independent components corresponding to its many functional parts. The design of each component can then be carried out with some degree of independence of the design of others, since each will affect the others largely through its function and independently of the details of the mechanisms that accomplish the function.[12]

There is no reason to expect that the decomposition of the complete design into functional components will be unique. In important instances there may exist alternative feasible decompositions of radically different kinds. This possibility is well known to designers of administrative organizations, where work can be divided up by subfunctions, by subprocesses, by subareas, and in other ways. Much of classical organization theory, in fact, was concerned precisely with this issue of alternative decompositions of a collection of interrelated tasks.

[11] I have developed this argument at greater length in my essay "The Architecture of Complexity," which follows this chapter.

[12] This approach to the design of complex structures has been explored by Christopher Alexander in *Notes on the Synthesis of Form* (Cambridge: Harvard University Press, 1967). He has also presented in his book some automated procedures for finding plausible decompositions once the matrix of interconnections of component functions has been specified.

The Generator-Test Cycle

One way of considering the decomposition, but acknowledging that the interrelations among the components cannot be ignored completely, is to think of the design process as involving first the generation of alternatives and then the testing of these alternatives against a whole array of requirements and constraints. There need not be merely a single generate-test cycle, but there can be a whole nested series of such cycles. The generators implicitly define the decomposition of the design problem, and the tests guarantee that important indirect consequences will be noticed and weighed. Alternative decompositions correspond to different ways of dividing the responsibilities for the final design between generators and tests.

To take a greatly oversimplified example, a series of generators may generate one or more possible outlines and schemes of fenestration for a building, while tests may be applied to determine whether needs for particular kinds of rooms can be met within the outlines generated. Alternatively, the generators may be used to evolve the structure of rooms, while tests are applied to see whether they are consistent with an acceptable over-all shape and design. The house can be designed from the outside in or from the inside out.[13]

Alternatives are also open, in organizing the design process, as to how far development of possible subsystems will be carried before the over-all coordinating design is developed in detail; or, vice versa, how far the over-all design should be carried before various components, or possible components, are developed. These alternatives of design are familiar to architects. They are familiar also to composers, who must decide how far the architectonics of a musical structure will be evolved before some of the component musical themes and other elements have been invented. Computer programmers face the same choices, between working downward from executive routines to

[13] I am indebted to John Grason for many ideas on the topic of this section. Mr. Grason is constructing a system for design by computer that will provide a means for further exploration of these matters.

subroutines or upward from component subroutines to a coordinating executive.

A theory of design will include principles—most of which do not yet exist—for deciding such questions of precedence and sequence in the design process.

Process as a Determinant of Style

When we recall that the process will generally be concerned with finding a satisfactory design, rather than an optimum design, we see that sequence and the division of labor between generators and tests can affect not only the efficiency with which resources for designing are used but also the nature of the final design as well. What we ordinarily call "style" may stem just as much from these decisions about the design process as from alternative emphasis on the goals to be realized through the final design. An architect who designs buildings from the outside in will arrive at quite different buildings from one who designs from the inside out, even though both of them might agree on the characteristics that a satisfactory building should possess.

When we come to the design of systems as complex as cities, or buildings, or economies, we must give up the aim of creating systems that will optimize some hypothesized utility function; and we must consider whether differences in style of the sort I have just been describing do not represent highly desirable variants in the design process rather than alternatives to be evaluated as "better" or "worse." Variety, within the limits of satisfactory constraints, may be a desirable end in itself, among other reasons, because it permits us to attach value to the search as well as its outcome—to regard the design process as itself a valued activity for those who participate in it.

We have usually thought of city planning as a means whereby the planner's creative activity could build a system that would satisfy the needs of a populace. Perhaps we should think of city planning as a valuable creative activity in which many members of a community can have the opportunity of participating—if we have wits to organize the process that way.

However that may be, I hope I have illustrated sufficiently that both the shape of the design and the shape and organization of the design process are essential components of a theory of design. These topics constitute the sixth item in my proposed curriculum in design:

6. *The organization of complex structures and its implication for the organization of design processes.*

Representation of the Design

I have by no means surveyed all facets of the emerging science of design. In particular, I have said little about the influence of problem representation on design. Although the importance of the question is recognized today, we have little systematic knowledge about it. I shall cite one example, to make clear what I mean by "representation."

Here are the rules of a game, which I shall call number scrabble. The game is played by two people with nine cards—let us say the ace through the nine of hearts. The cards are placed in a row, face up, between the two players. The players draw alternately, one at a time, selecting any one of the cards that remain in the center. The aim of the game is for a player to make up a "book," that is, a set of exactly three cards whose spots add to 15, before his opponent can do so. The first player who makes a book wins; if all nine cards have been drawn without either player making a book, the game is a draw.

What is a good strategy in this game? How would you go about finding one? If the reader has not already discovered it for himself, let me show how a change in representation will make it easy to play the game well. The magic square here, which I introduced in the second chapter, is made up of the numerals from 1 through 9.

4	9	2
3	5	7
8	1	6

Each row, column, or diagonal adds to 15; and every triple of these numerals that adds to 15 is a row, column, or diagonal of the magic square. From this, it is obvious that "making a book" in number scrabble is equivalent to getting "three in a row" in the game of tic-tac-toe. But most people know how to play tic-tac-toe well, hence can simply transfer their usual strategy to number scrabble.[14]

Problem Solving as Change in Representation

That representation makes a difference is a long-familiar point. We all believe that arithmetic has become easier since Arabic numerals and place notation replaced Roman numerals, although I know of no theoretic treatment that explains why.[15]

That representation makes a difference is evident for a different reason. All mathematics exhibits in its conclusions only what is already implicit in its premises, as I mentioned in a previous chapter. Hence all mathematical derivation can be viewed simply as change in representation, making evident what was previously true but obscure.

This view can be extended to all of problem solving—solving a problem simply means representing it so as to make the solution transparent.[16] If the problem solving could actually be organized in these terms, the issue of representation would indeed become central. But even if it cannot—if this is too exaggerated a view—a deeper understanding of how representations are created and how they contribute to the solution of problems will become an essential component in the future theory of design.

Spatial Representation

Since much of design, particularly architectural and engineering design, is concerned with objects or arrangements

[14] Number scrabble is not the only isomorph of tic-tac-toe. John A. Michon has described another, JAM, which is the dual of tic-tac-toe in the sense of projective geometry. That is, the rows, columns, and diagonals of tic-tac-toe become points in JAM, and the squares of the former become line segments joining the points. The game is won by "jamming" all the segments through a point—a move consists of seizing or jamming a single segment. Other isomorphs of tic-tac-toe are known as well.

[15] My colleague, Allen Newell, has been investigating this question. I shall not try to anticipate his answer.

[16] Saul Amarel, "On the Mechanization of Creative Processes," IEEE *Spectrum 3* (No. 4): 112–114 (April 1966).

in real Euclidean two-dimensional or three-dimensional space, the representation of space and of things in space will necessarily be a central topic in a science of design. From our previous discussion of visual perception, it should be clear that "space" inside the head of the designer or the memory of a computer may have very different properties from a picture on paper or a three-dimensional model.

These representational issues have already attracted the attention of those concerned with computer-aided design—the cooperation of human and computer in the design process. As a single example, I may mention Ivan Sutherland's SKETCHPAD program, which allows geometric shapes to be represented and conditions to be placed on these shapes in terms of constraints, to which they then conform.[17]

Geometric considerations are also prominent in the few attempts made thus far to automate completely the design, say, of printed or etched circuits, or of buildings. Grason, for example, in a system for designing house floor plans, constructs an internal representation of the layout that helps one decide whether a proposed set of connections among rooms, selected to meet design criteria for communication, and so on, can be realized in a plane (private communication).

The Taxonomy of Representation

An early step toward understanding any set of phenomena is to learn what kinds of things there are in the set—to develop a taxonomy. This step has not yet been taken with respect to representations. We have only a sketchy and incomplete knowledge of the different ways in which problems can be represented and much less knowledge of the significance of the differences.

In a completely pragmatic vein, we know that problems can be described verbally, in natural language. They often can be described mathematically, using standard formalisms of algebra, geometry, set theory, analysis, or topology.

[17] I. E. Sutherland, "SKETCHPAD, A Man-Machine Graphical Communication System," *Proceedings, AFIPS Spring Joint Computer Conference, 1963* (Baltimore: Spartan Books), pp. 329–346.

If the problems relate to physical objects, they (or their solutions) can be represented by floor plans, engineering drawings, renderings, or three-dimensional models. Problems that have to do with actions can be attacked with flow charts and programs.

Other items most likely will need to be added to the list, and there may exist more fundamental and significant ways of classifying its members. But even though our classification is incomplete, and perhaps superficial, we can begin to build a theory of the properties of these representations. A number of topics in the growing theories of machines and of programming languages may give us some notion of the directions that a theory of representations—at least on its more formal side—may take.[18] These topics can also provide, at the beginning, some of the substance for the final subject in our program on the theory of design:

7. *Alternative representations for design problems.*

Summary—Topics in the Theory of Design

My main goal in this chapter has been to show that there already exist today a number of components of a theory of design and a substantial body of knowledge, theoretical and empirical, relating to each. As we draw up our curriculum in design—in the science of the artificial—to take its place by the side of natural science in the whole engineering curriculum, it includes at least the following topics:

THE EVALUATION OF DESIGNS

1. Theory of evaluation: utility theory, statistical decision theory
2. Computational methods:
   a. Algorithms for choosing *optimal* alternatives such as linear programming computations, control theory, dynamic programming
   b. Algorithms and heuristics for choosing *satisfactory* alternatives

[18] By way of example, see Marvin L. Minsky, *Computation: Finite and Infinite Machines* (Englewood Cliffs, N. J.: Prentice-Hall, 1967), and Kenneth E. Iverson, *A Programming Language* (New York: Wiley, 1962).

3. THE FORMAL LOGIC OF DESIGN: Imperative and declarative logics

THE SEARCH FOR ALTERNATIVES

4. Heuristic search: factorization and means-end analysis
5. Allocation of resources for search

6. THEORY OF STRUCTURE AND DESIGN ORGANIZATION: Hierarchic systems
7. REPRESENTATION OF DESIGN PROBLEMS

In small segments of the curriculum—the theory of evaluation, for example, and the formal logic of design—it is already possible to organize the instruction within a framework of systematic, formal theory. In many other segments the treatment would be more pragmatic, more empirical.

But nowhere do we need to return or retreat to the methods of the cookbook that originally put design into disrepute and drove it from the engineering curriculum. For there exist today a considerable number of examples of actual design processes, of many different kinds, that have been defined fully and cast in the metal, so to speak, in the form of running computer programs: optimizing algorithms, search procedures, and special-purpose programs for designing motors, balancing assembly lines, selecting investment portfolios, locating warehouses, designing highways, and so forth.

Because these computer programs describe complex design processes in complete, painstaking detail, they are open to full inspection and analysis, or to trial by simulation. They constitute a body of empirical phenomena to which the student of design can address himself and which he can seek to understand. There is no question, since these programs exist, of the design process hiding behind the cloak of "judgment" or "experience." Whatever judgment or experience was used in creating the programs must now be incorporated in them and hence be observable. The programs are the tangible record of the variety of schemes that man has devised to explore his complex outer environment, and to discover in that environment the paths to his goals.

ROLE OF DESIGN IN THE LIFE OF THE MIND

I have called my topic "the theory of design," and my curriculum a "program in design." I have emphasized its role as complement to the natural science curriculum in the total training of a professional engineer—or of any professional whose task is to solve problems, to choose, to synthesize, to decide.

But there is another way in which the theory of design may be viewed in relation to other knowledge. My second chapter was a chapter on psychology—specifically on man's relation to his biological inner environment. The present chapter may also be construed as a chapter on psychology: on man's relation to the complex outer environment in which he seeks to survive and achieve.

Both chapters, so construed, have significance that goes beyond the professional work of the man we have called the "designer." Many of us have been unhappy about the fragmentation of our society into two cultures. Some of us even think there are not just two cultures but a large number of cultures. If we regret that fragmentation, then we must look for a common core of knowledge that can be shared by the members of all cultures—a core that includes more significant topics than the weather, sports, automobiles, the care and feeding of children, or perhaps even politics. A common understanding of our relation to the inner and outer environments that define the space in which we live and choose can provide at least part of that significant core.

This may seem an extravagant claim. Let me use the realm of music to illustrate what I mean. Music is one of the most ancient of the sciences of the artificial, and was so recognized by the Greeks. Anything I have said about the artificial would apply as well to music, its composition or its enjoyment, as to the engineering topics I have used for most of my illustrations.

Music involves a formal pattern. It has few (but important) contacts with the inner environment; that is, it is capable of evoking strong emotions, its patterns are detectable by human listeners, and some of its harmonic relations can be given physical and physiological interpretations (though the aesthetic import of these is debatable). As for the outer environment, when we view composition as a problem in design, we encounter just the same tasks of evaluation, of search for alternatives, and of representation that we do in any other design problem. If it pleases us, we can even apply to music some of the same techniques of automatic design by computer that have been used in other fields of design. If computer-composed music has not yet reached notable heights of aesthetic excellence, it deserves, and has already received, serious attention from professional composers and analysts, who do not find it written in tongues alien to them.[19]

Undoubtedly, there are tone-deaf engineers, just as there are mathematically ignorant composers. Few engineers and composers, whether deaf, ignorant, or not, can carry on a mutually rewarding conversation about the content of each other's professional work. What I am suggesting is that they *can* carry on such a conversation about design, can begin to perceive the common creative activity in which they are both engaged, can begin to share their experiences of the creative, professional design process.

Those of us who have lived close to the development of the modern computer through gestation and infancy have been drawn from a wide variety of professional fields, music being one of them. We have noticed the growing communication among intellectual disciplines that takes place around the computer. We have welcomed it, because it has brought us into contact with new worlds of knowledge—has helped us combat our own multiple-cultures isolation. This breakdown of old disciplinary boundaries has been much commented upon, and its connection with computers and the information sciences often noted.

But surely the computer, as a piece of hardware, or even as a piece of programmed software, has nothing to do, directly, with the matter. I have already suggested a different explanation. The ability to communicate across

[19] L. A. Hillier and L. M. Isaacson's *Experimental Music* (New York: McGraw-Hill, 1959), reporting experiments begun more than a decade ago, still provides a good introduction to the subject of musical composition, viewed as design. See also Walter R. Reitman, *Cognition and Thought* (New York: Wiley, 1965), Chapter 6, "Creative Problem Solving: Notes from the Autobiography of a Fugue."

fields—the common ground—comes from the fact that all who use computers in complex ways are using computers to design, or to participate in the process of design. Consequently, we as designers, or as designers of design processes, have had to be explicit, as never before, about what is involved in creating a design and what takes place while the creation is going on.

The real subjects of the new intellectual free trade among the many cultures are our own thought processes, our processes of judging, deciding, choosing, and creating. We are importing and exporting from one intellectual discipline to another ideas about how a serially organized information-processing system like a man—or a computer, or a complex of men and computers in organized co-operation—solves problems and achieves goals in outer environments of great complexity.

The proper study of mankind has been said to be man. But I have argued that man—or at least the intellective component of man—may be relatively simple; that most of the complexity of his behavior may be drawn from his environment, from his search for good designs. If I have made my case, then we can conclude that, in large part, the proper study of mankind is the science of design, not only as the professional component of a technical education but as a core discipline for every liberally educated man.

4 THE ARCHITECTURE OF COMPLEXITY*

A NUMBER of proposals have been advanced in recent years for the development of "general systems theory" that, abstracting from properties peculiar to physical, biological, or social systems, would be applicable to all of them.[1] We might well feel that,

* Reprinted with permission from *Proceedings of the American Philosophical Society*, *106*: 467–482 (December 1962).

[1] See especially the yearbooks of the Society for General Systems Research. Prominent among the exponents of general systems theory are L. von Bertalanffy, K. Boulding, R. W. Gerard, and J. G. Miller. For a more skeptical view—perhaps too skeptical in the light of the present discussion—see H. A. Simon and A. Newell, "Models: Their Uses and Limitations," in L. D. White (ed.), *The State of the Social Sciences* (Chicago: University of Chicago Press, 1956), pp. 66–83.

while the goal is laudable, systems of such diverse kinds could hardly be expected to have any nontrivial properties in common. Metaphor and analogy can be helpful, or they can be misleading. All depends on whether the similarities the metaphor captures are significant or superficial.

It may not be entirely vain, however, to search for common properties among diverse kinds of complex systems. The ideas that go by the name of cybernetics constitute, if not a theory, at least a point of view that has been proving fruitful over a wide range of applications.[2] It has been useful to look at the behavior of adaptive systems in terms of the concepts of feedback and homeostasis, and to analyze adaptiveness in terms of the theory of selective information.[3] The ideas of feedback and information provide a frame of reference for viewing a wide range of situations, just as do the ideas of evolution, or relativism, of axiomatic method, and of operationalism.

In this essay I should like to report on some things we have been learning about particular kinds of complex systems encountered in the behavioral sciences. The developments I shall discuss arose in the context of specific phenomena, but the theoretical formulations themselves make little reference to details of structure. Instead they refer primarily to the complexity of the systems under view without specifying the exact content of that complexity. Because of their abstractness, the theories may have relevance—application would be too strong a term—to other kinds of complex systems that are observed in the social, biological, and physical sciences.

In recounting these developments, I shall avoid technical detail, which can generally be found elsewhere. I shall describe each theory in the particular context in which it arose. Then, I shall cite some examples of complex systems, from areas of science other than the initial application, to which the theoretical framework appears relevant. In doing so, I shall make reference to areas of knowledge

[2] N. Wiener, *Cybernetics* (New York: Wiley, 1948). For an imaginative forerunner, see A. J. Lotka, *Elements of Mathematical Biology* (New York: Dover Publications, 1951), first published in 1924 as *Elements of Physical Biology*.

[3] C. Shannon and W. Weaver, *The Mathematical Theory of Communication* (Urbana: University of Illinois Press, 1949); W. R. Ashby, *Design for a Brain* (New York: Wiley, 1952).

where I am not expert—perhaps not even literate. The reader will have little difficulty, I am sure, in distinguishing instances based on idle fancy or sheer ignorance from instances that cast some light on the ways in which complexity exhibits itself wherever it is found in nature.

I shall not undertake a formal definition of "complex systems."[4] Roughly, by a complex system I mean one made up of a large number of parts that interact in a nonsimple way. In such systems, the whole is more than the sum of the parts, not in an ultimate, metaphysical sense, but in the important pragmatic sense that, given the properties of the parts and the laws of their interaction, it is not a trivial matter to infer the properties of the whole. In the face of complexity, an in-principle reductionist may be at the same time a pragmatic holist.[5]

The four sections that follow discuss four aspects of complexity. The first offers some comments on the frequency with which complexity takes the form of hierarchy —the complex system being composed of subsystems that, in turn, have their own subsystems, and so on. The second section theorizes about the relation between the structure of a complex system and the time required for it to emerge through evolutionary processes; specifically, it argues that hierarchic systems will evolve far more quickly than nonhierarchic systems of comparable size. The third section explores the dynamic properties of hierarchically organized systems and shows how they can be decomposed into subsystems in order to analyze their behavior. The fourth section examines the relation between complex systems and their descriptions.

[4] W. Weaver, in "Science and Complexity," *American Scientist*, *36*: 536 (1948), has distinguished two kinds of complexity, disorganized and organized. We shall be concerned primarily with organized complexity.

[5] See also John R. Platt, "Properties of Large Molecules That Go beyond the Properties of Their Chemical Sub-groups," *Journal of Theoretical Biology*, *1*: 342–358 (1961). Since the reductionism-holism issue is a major *cause de guerre* between scientists and humanists, perhaps we might even hope that peace could be negotiated between the two cultures along the lines of the compromise just suggested. As I go along, I shall have a little to say about complexity in the arts as well as in the natural sciences. I must emphasize the pragmatism of my holism to distinguish it sharply from the position taken by W. M. Elsasser in *The Physical Foundation of Biology* (New York: Pergamon Press, 1958).

Thus, my central theme is that complexity frequently takes the form of hierarchy and that hierarchic systems have some common properties that are independent of their specific content. Hierarchy, I shall argue, is one of the central structural schemes that the architect of complexity uses.

Hierarchic systems

By a *hierarchic system*, or hierarchy, I mean a system that is composed of interrelated subsystems, each of the latter being, in turn, hierarchic in structure until we reach some lowest level of elementary subsystem. In most systems in nature, it is somewhat arbitrary as to where we leave off the partitioning and what subsystems we take as elementary. Physics makes much use of the concept of "elementary particle," although particles have a disconcerting tendency not to remain elementary very long. Only a couple of generations ago, the atoms themselves were elementary particles; today, to the nuclear physicist they are complex systems. For certain purposes of astronomy, whole stars, or even galaxies, can be regarded as elementary subsystems. In one kind of biological research, a cell may be treated as an elementary subsystem; in another, a protein molecule; in still another, an amino acid residue.

Just why a scientist has a right to treat as elementary a subsystem that is in fact exceedingly complex is one of the questions we shall take up. For the moment, we shall accept the fact that scientists do this all the time and that, if they are careful scientists, they usually get away with it.

Etymologically, the word "hierarchy" has had a narrower meaning than I am giving it here. The term has generally been used to refer to a complex system in which each of the subsystems is subordinated by an authority relation to the system it belongs to. More exactly, in a hierarchic formal organization, each system consists of a "boss" and a set of subordinate subsystems. Each of the subsystems has a "boss" who is the immediate subordinate of the boss of the system. We shall want to consider systems in which the relations among subsystems are more complex than in the formal organizational hierarchy just described. We shall want to include systems in which there is no relation of subordination among subsystems. (In fact, even in human organizations, the formal hierarchy exists only on paper; the real flesh-and-blood organization has many interpart relations other than the lines of formal authority.) For lack of a better term, I shall use "hierarchy" in the broader sense introduced in the previous paragraphs, to refer to all complex systems analyzable into successive sets of subsystems, and speak of "formal hierarchy" when I want to refer to the more specialized concept.[6]

Social Systems

I have already given an example of one kind of hierarchy that is frequently encountered in the social sciences: a formal organization. Business firms, governments, and universities all have a clearly visible parts-within-parts structure. But formal organizations are not the only, or even the most common, kind of social hierarchy. Almost all societies have elementary units called families, which may be grouped into villages or tribes, and these into larger groupings, and so on. If we make a chart of social interactions, of who talks to whom, the clusters of dense interaction in the chart will identify a rather well-defined hierarchic structure. The groupings in this structure may be defined operationally by some measure of frequency of interaction in this sociometric matrix.

Biological and Physical Systems

The hierarchical structure of biological systems is a familiar fact. Taking the cell as the building block, we find cells organized into tissues, tissues into organs, organs into systems. Moving downward from the cell, well-defined subsystems—for example, nucleus, cell membrane, microsomes, mitochondria, and so on—have been identified in animal cells.

[6] The mathematical term "partitioning" will not do for what I call here a hierarchy; for the set of subsystems, and the successive subsets in each of these defines the partitioning, independently of any systems of relations among the subsets. By "hierarchy" I mean the partitioning in conjunction with the relations that hold among its parts.

The hierarchic structure of many physical systems is equally clear-cut. I have already mentioned the two main series. At the microscopic level we have elementary particles, atoms, molecules, and macromolecules. At the macroscopic level we have satellite systems, planetary systems, galaxies. Matter is distributed throughout space in a strikingly nonuniform fashion. The most nearly random distributions we find, gases, are not random distributions of elementary particles but random distributions of complex systems, that is, molecules.

A considerable range of structural types is subsumed under the term "hierarchy" as I have defined it. By this definition, a diamond is hierarchic, for it is a crystal structure of carbon atoms that can be further decomposed into protons, neutrons, and electrons. However, it is a very "flat" hierarchy, in which the number of first-order subsystems belonging to the crystal can be indefinitely large. A volume of molecular gas is a flat hierarchy in the same sense. In ordinary usage, we tend to reserve the word "hierarchy" for a system that is divided into a *small or moderate number* of subsystems, each of which may be further subdivided. Hence, we do not ordinarily think of or refer to a diamond or a gas as a hierarchic structure. Similarly, a linear polymer is simply a chain, which may be very long, of identical subparts, the monomers. At the molecular level it is a very flat hierarchy.

In discussing formal organizations, the number of subordinates who report directly to a single boss is called his *span of control*. I shall speak analogously of the *span* of a system, by which I shall mean the number of subsystems into which it is partitioned. Thus, a hierarchic system is flat at a given level if it has a wide span at that level. A diamond has a wide span at the crystal level, but not at the next level down, the molecular level.

In most of our theory construction in the following sections we shall focus our attention on hierarchies of moderate span, but from time to time I shall comment on the extent to which the theories might or might not be expected to apply to very flat hierarchies.

There is one important difference between the physical and biological hierarchies, on the one hand, and social hierarchies, on the other. Most physical and biological hierarchies are described in spatial terms. We detect the organelles in a cell in the way we detect the raisins in a cake—they are "visibly" differentiated substructures localized spatially in the larger structure. On the other hand, we propose to identify social hierarchies not by observing who lives close to whom but by observing who interacts with whom. These two points of view can be reconciled by defining hierarchy in terms of intensity of interaction, but observing that in most biological and physical systems relatively intense interaction implies relative spatial propinquity. One of the interesting characteristics of nerve cells and telephone wires is that they permit very specific strong interactions at great distances. To the extent that interactions are channeled through specialized communications and transportation systems, spatial propinquity becomes less determinative of structure.

Symbolic Systems

One very important class of systems has been omitted from my examples thus far: systems of human symbolic production. A book is a hierarchy in the sense in which I am using that term. It is generally divided into chapters, the chapters into sections, the sections into paragraphs, the paragraphs into sentences, the sentences into clauses and phrases, the clauses and phrases into words. We may take the words as our elementary units, or further subdivide them, as the linguist often does, into smaller units. If the book is narrative in character, it may divide into "episodes" instead of sections, but divisions there will be.

The hierarchic structure of music, based on such units as movements, parts, themes, phrases, is well known. The hierarchic structure of products of the pictorial arts is more difficult to characterize, but I shall have something to say about it later.

The Evolution of Complex Systems

Let me introduce the topic of evolution with a parable. There once were two watchmakers, named Hora and Tempus, who manufactured very fine watches. Both of them were highly regarded, and the phones in their workshops rang frequently—new customers were constantly calling them. However, Hora prospered, while Tempus became poorer and poorer and finally lost his shop. What was the reason?

The watches the men made consisted of about 1,000 parts each. Tempus had so constructed his that if he had one partly assembled and had to put it down—to answer the phone, say—it immediately fell to pieces and had to be reassembled from the elements. The better the customers liked his watches, the more they phoned him and the more difficult it became for him to find enough uninterrupted time to finish a watch.

The watches that Hora made were no less complex than those of Tempus. But he had designed them so that he could put together subassemblies of about ten elements each. Ten of these subassemblies, again, could be put together into a larger subassembly; and a system of ten of the latter subassemblies constituted the whole watch. Hence, when Hora had to put down a partly assembled watch in order to answer the phone, he lost only a small part of his work, and he assembled his watches in only a fraction of the man-hours it took Tempus.

It is rather easy to make a quantitative analysis of the relative difficulty of the tasks of Tempus and Hora: Suppose the probability that an interruption will occur while a part is being added to an incomplete assembly is p. Then the probability that Tempus can complete a watch he has started without interruption is $(1 - p)^{1000}$—a very small number unless p is 0.001 or less. Each interruption will cost, on the average, the time to assemble $1/p$ parts (the expected number assembled before interruption). On the other hand, Hora has to complete 111 subassemblies of ten parts each. The probability that he will not be interrupted while completing any one of these is $(1 - p)^{10}$, and each interruption will cost only about the time required to assemble five parts.[7]

[7] The speculations on speed of evolution were first suggested by H. Jacobson's application of information theory to estimating the time required for biological evolution. See his paper "Information, Reproduction, and the Origin of Life," in *American Scientist*, *43*: 119–127 (January 1955). From thermodynamic considerations it is possible to estimate the amount of increase in entropy that occurs

Now if p is about 0.01—that is, there is one chance in a hundred that either watchmaker will be interrupted while adding any one part to an assembly—then a straightforward calculation shows that it will take Tempus, on the average, about four thousand times as long to assemble a watch as Hora.

We arrive at the estimate as follows:

1. Hora must make 111 times as many complete assemblies per watch as Tempus; but
2. Tempus will lose on the average 20 times as much work for each interrupted assembly as Hora (100 parts, on the average, as against 5); and
3. Tempus will complete an assembly only 44 times per million attempts ($0.99^{1000} = 44 \times 10^{-6}$), while Hora will complete nine out of ten ($0.99^{10} = 9 \times 10^{-1}$). Hence Tempus will have to make 20,000 as many attempts per completed assembly as Hora. $(9 \times 10^{-1})/(44 \times 10^{-6}) = 2 \times 10^4$. Multiplying these three ratios, we get

$$1/111 \times 100/5 \times 0.99^{10}/0.99^{1000}$$
$$= 1/111 \times 20 \times 20{,}000 \sim 4{,}000.$$

Biological Evolution

What lessons can we draw from our parable for biological evolution? Let us interpret a partially completed subassembly of k elementary parts as the coexistence of k parts in a small volume—ignoring their relative orientations.

---

when a complex system decomposes into its elements. (See for example, R. B. Setlow and E. C. Pollard, *Molecular Biophysics* (Reading, Mass.: Addison-Wesley, 1962), pp. 63–65, and references cited there.) But entropy is the logarithm of a probability, hence information, the negative of entropy, can be interpreted as the logarithm of the reciprocal of the probability—the "improbability," so to speak. The essential idea in Jacobson's model is that the expected time required for the system to reach a particular state is inversely proportional to the probability of the state—hence it increases exponentially with the amount of information (negentropy) of the state.

Following this line of argument, but not introducing the notion of levels and stable subassemblies, Jacobson arrived at estimates of the time required for evolution so large as to make the event rather improbable. Our analysis, carried through in the same way, but with attention to the stable intermediate forms, produces very much smaller estimates.

The model assumes that parts are entering the volume at a constant rate, but that there is a constant probability, p, that the part will be dispersed before another is added, unless the assembly reaches a stable state. These assumptions are not particularly realistic. They undoubtedly underestimate the decrease in probability of achieving the assembly with increase in the size of the assembly. Hence the assumptions understate—probably by a large factor—the relative advantage of a hierarchic structure.

Although we cannot, therefore, take the numerical estimate seriously, the lesson for biological evolution is quite clear and direct. The time required for the evolution of a complex form from simple elements depends critically on the numbers and distribution of potential intermediate stable forms. In particular, if there exists a hierarchy of potential stable "subassemblies," with about the same span, s, at each level of the hierarchy, then the time required for a subassembly can be expected to be about the same at each level—that is, proportional to $1/(1 - p)^s$. The time required for the assembly of a system of n elements will be proportional to $\log_s n$, that is, to the number of levels in the system. One would say—with more illustrative than literal intent—that the time required for the evolution of multicelled organisms from single-celled organisms might be of the same order of magnitude as the time required for the evolution of single-celled organisms from macromolecules. The same argument could be applied to the evolution of proteins from amino acids, of molecules from atoms, of atoms from elementary particles.

A whole host of objections to this oversimplified scheme will occur, I am sure, to every working biologist, chemist, and physicist. Before turning to matters I know more about, I shall mention three of these problems, leaving the rest to the attention of the specialists.

First, in spite of the overtones of the watchmaker parable, the theory assumes no teleological mechanism. The complex forms can arise from the simple ones by purely random processes. (I shall propose another model in a moment that shows this clearly.) Direction is provided to the scheme by the stability of the complex forms, once these come into existence. But this is nothing more than survival of the fittest—that is, of the stable.

Second, not all large systems appear hierarchical. For example, most polymers—such as nylon—are simply linear chains of large numbers of identical components, the monomers. However, for present purposes we can simply regard such a structure as a hierarchy with a span of one—the limiting case. For a chain of any length represents a state of relative equilibrium.[8]

Third, the evolution of complex systems from simple elements implies nothing, one way or the other, about the change in entropy of the entire system. If the process absorbs free energy, the complex system will have a smaller entropy than the elements; if it releases free energy, the opposite will be true. The former alternative is the one that holds for most biological systems, and the net inflow of free energy has to be supplied from the sun or some other source if the second law of thermodynamics is not to be violated. For the evolutionary process we are describing, the equilibria of the intermediate states need have only local and not global stability, and they may be stable only in the steady state—that is, as long as there is an external source of free energy that may be drawn upon.[9]

Because organisms are not energetically closed systems, there is no way to deduce the direction, much less the rate, of evolution from classical thermodynamic considerations. All estimates indicate that the amount of entropy, measured in physical units, involved in the formation of a one-celled biological organism is trivially small—about -10^{-11} cal/degree.[10] The "improbability" of evolution

[8] There is a well-developed theory of polymer size, based on models of random assembly. See, for example, P. J. Flory, *Principles of Polymer Chemistry* (Ithaca: Cornell University Press, 1953), Chapter 8. Since *all* subassemblies in the polymerization theory are stable, limitation of molecular growth depends on "poisoning" of terminal groups by impurities or formation of cycles rather than upon disruption of partially formed chains.

[9] This point has been made many times before, but it cannot be emphasized too strongly. For further discussion, see Setlow and Pollard, *op cit.*, pp. 49–64; E. Schrödinger, *What Is Life?* (Cambridge: Cambridge University Press, 1945); and H. Linschitz, "The Information Content of a Bacterial Cell," in H. Quastler (ed.), *Information Theory in Biology* (Urbana: University of Illinois Press, 1953), pp. 251–262.

[10] See Linschitz, *op. cit.* This quantity, 10^{-11} cal/degree, corresponds to about 10^{13} bits of information.

has nothing to do with this quantity of entropy, which is produced by every bacterial cell every generation. The irrelevance of quantity of information, in this sense, to speed of evolution can also be seen from the fact that exactly as much information is required to "copy" a cell through the reproductive process as to produce the first cell through evolution.

The effect of the existence of stable intermediate forms exercises a powerful effect on the evolution of complex forms that may be likened to the dramatic effect of catalysts upon reaction rates and steady-state distribution of reaction products in open systems.[11] In neither case does the entropy change provide us with a guide to system behavior.

Problem Solving as Natural Selection

Let us turn now to some phenomena that have no obvious connection with biological evolution: human problem-solving processes. Consider, for example, the task of discovering the proof for a difficult theorem. The process can be—and often has been—described as a search through a maze. Starting with the axioms and previously proved theorems, various transformations allowed by the rules of the mathematical systems are attempted, to obtain new expressions. These are modified in turn until, with persistence and good fortune, a sequence or path of transformations is discovered that leads to the goal.

The process ordinarily involves much trial and error. Various paths are tried; some are abandoned, others are pushed further. Before a solution is found, many paths of the maze may be explored. The more difficult and novel the problem, the greater is likely to be the amount of trial and error required to find a solution. At the same time, the trial and error is not completely random or blind; it is, in fact, rather highly selective. The new expressions that are obtained by transforming given ones are examined to see whether they represent progress toward the goal. Indications of progress spur further search in

[11] See H. Kacser, "Some Physico-chemical Aspects of Biological Organization," Appendix, pp. 191–249, in C. H. Waddington, *The Strategy of the Genes* (London: George Allen & Unwin, 1957).

the same direction; lack of progress signals the abandonment of a line of search. Problem solving requires *selective* trial and error.[12]

A little reflection reveals that cues signaling progress play the same role in the problem-solving process that stable intermediate forms play in the biological evolutionary process. In fact, we can take over the watchmaker parable and apply it also to problem solving. In problem solving, a partial result that represents recognizable progress toward the goal plays the role of a stable subassembly.

Suppose that the task is to open a safe whose lock has 10 dials, each with 100 possible settings, numbered from 0 to 99. How long will it take to open the safe by a blind trial-and-error search for the correct setting? Since there are 100^{10} possible settings, we may expect to examine about one half of these, on the average, before finding the correct one—that is, 50 billion billion settings. Suppose, however, that the safe is defective, so that a click can be heard when any one dial is turned to the correct setting. Now each dial can be adjusted independently and does not need to be touched again while the others are being set. The total number of settings that have to be tried is only 10×50, or 500. The task of opening the safe has been altered, by the cues the clicks provide, from a practically impossible one to a trivial one.[13]

[12] See A. Newell, J. C. Shaw, and H. A. Simon, "Empirical Explorations of the Logic Theory Machine," *Proceedings of the 1957 Western Joint Computer Conference*, February 1957 (New York: Institute of Radio Engineers); "Chess-Playing Programs and the Problem of Complexity," *IBM Journal of Research and Development*, *2*: 320–335 (October 1958); and for a similar view of problem solving, W. R. Ashby, "Design for an Intelligence Amplifier," pp. 215–233 in C. E. Shannon and J. McCarthy, *Automata Studies* (Princeton: Princeton University Press, 1956).

[13] The clicking safe example was supplied by D. P. Simon. Ashby, *op. cit.*, p. 230, has called the selectivity involved in situations of this kind "selection by components." The even greater reduction in time produced by hierarchization in the clicking safe example, as compared with the watchmaker's metaphor, is due to the fact that a random *search* for the correct combination is involved in the former case, while in the latter the parts come together in the right order. It is not clear which of these metaphors provides the better model for biological evolution, but we may be sure that the watchmaker's metaphor gives an exceedingly conservative estimate of the savings due to hierarchization. The safe may give an excessively high estimate because it assumes all possible arrangements of the elements to be equally probable.

A considerable amount has been learned in the past five years about the nature of the mazes that represent common human problem-solving tasks—proving theorems, solving puzzles, playing chess, making investments, balancing assembly lines, to mention a few. All that we have learned about these mazes points to the same conclusion: that human problem solving, from the most blundering to the most insightful, involves nothing more than varying mixtures of trial and error and selectivity. The selectivity derives from various rules of thumb, or heuristics, that suggest which paths should be tried first and which leads are promising. We do not need to postulate processes more sophisticated than those involved in organic evolution to explain how enormous problem mazes are cut down to quite reasonable size.[14]

The Sources of Selectivity

When we examine the sources from which the problem-solving system, or the evolving system, as the case may be, derives its selectivity, we discover that selectivity can always be equated with some kind of feedback of information from the environment.

Let us consider the case of problem solving first. There are two basic kinds of selectivity. One we have already noted: various paths are tried out, the consequences of following them are noted, and this information is used to guide further search. In the same way, in organic evolution, various complexes come into being, at least evanescently, and those that are stable provide new building blocks for further construction. It is this information about stable configurations, and not free energy or negentropy from the sun, that guides the process of evolution and provides the selectivity that is essential to account for its rapidity.

The second source of selectivity in problem solving is previous experience. We see this particularly clearly when the problem to be solved is similar to one that has been solved before. Then, by simply trying again the paths that

[14] A. Newell and H. A. Simon, "Computer Simulation of Human Thinking," *Science, 134*: 2011–2017 (December 22, 1961).

led to the earlier solution, or their analogues, trial-and-error search is greatly reduced or altogether eliminated.

What corresponds to this latter kind of information in organic evolution? The closest analogue is reproduction. Once we reach the level of self-reproducing systems, a complex system, when it has once been achieved, can be multiplied indefinitely. Reproduction, in fact, allows the inheritance of acquired characteristics, but at the level of genetic material, of course; that is, only characteristics acquired by the genes can be inherited. We shall return to the topic of reproduction in the final section of this essay.

On Empires and Empire Building

We have not exhausted the categories of complex systems to which the watchmaker argument can reasonably be applied. Philip assembled his Macedonian empire and gave it to his son, to be later combined with the Persian subassembly and others into Alexander's greater system. On Alexander's death, his empire did not crumble to dust but fragmented into some of the major subsystems that had composed it.

The watchmaker argument implies that if one would be Alexander, one should be born into a world where large stable political systems already exist. Where this condition was not fulfilled, as on the Scythian and Indian frontiers, Alexander found empire building a slippery business. So too, T. E. Lawrence's organizing of the Arabian revolt against the Turks was limited by the character of his largest stable building blocks, the separate, suspicious desert tribes.

The profession of history places a greater value upon the validated particular fact than upon tendentious generalization. I shall not elaborate upon my fancy, therefore, but shall leave it to historians to decide whether anything can be learned for the interpretation of history from an abstract theory of hierarchic complex systems.

Conclusion: The Evolutionary Explanation of Hierarchy

We have shown thus far that complex systems will evolve from simple systems much more rapidly if there are stable intermediate forms than if there are not. The resulting complex forms in the former case will be hierarchic. We have only to turn the argument around to explain the observed predominance of hierarchies among the complex systems nature presents to us. Among possible complex forms, hierarchies are the ones that have the time to evolve. The hypothesis that complexity will be hierarchic makes no distinction among very flat hierarchies, like crystals and tissues and polymers, and the intermediate forms. Indeed, in the complex systems we encounter in nature, examples of both forms are prominent. A more complete theory than the one we have developed here would presumably have something to say about the determinants of width of span in these systems.

NEARLY DECOMPOSABLE SYSTEMS

In hierarchic systems, we can distinguish between the interactions *among* subsystems, on the one hand, and the interactions *within* subsystems—that is, among the parts of those subsystems—on the other. The interactions at the different levels may be, and often will be, of different orders of magnitude. In a formal organization there will generally be more interaction, on the average, between two employees who are members of the same department than between two employees from different departments. In organic substances, intermolecular forces will generally be weaker than molecular forces, and molecular forces weaker than nuclear forces.

In a rare gas, the intermolecular forces will be negligible compared to those binding the molecules—we can treat the individual particles, for many purposes, as if they were independent of each other. We can describe such a system as *decomposable* into the subsystems comprised of the individual particles. As the gas becomes denser, molecular interactions become more significant. But over some range, we can treat the decomposable case as a limit and as a first approximation. We can use a theory of perfect gases, for example, to describe approximately the behavior of actual gases if they are not too dense. As a second approximation, we may move to a theory of *nearly decomposable* systems, in which the interactions among the subsystems are weak but not negligible.

At least some kinds of hierarchic systems can be approximated successfully as nearly decomposable systems. The main theoretical findings from the approach can be summed up in two propositions: (*a*) in a nearly decomposable system, the short-run behavior of each of the component subsystems is approximately independent of the short-run behavior of the other components; (*b*) in the long run, the behavior of any one of the components depends in only an aggregate way on the behavior of the other components.

Let me provide a very concrete simple example of a nearly decomposable system.[15] Consider a building whose outside walls provide perfect thermal insulation from the environment. We shall take these walls as the boundary of our system. The building is divided into a large number of rooms, the walls between them being good, but not perfect, insulators. The walls between rooms are the boundaries of our major subsystems. Each room is divided by partitions into a number of cubicles, but the partitions are poor insulators. A thermometer hangs in each cubicle. Suppose that at the time of our first observation of the system there is a wide variation in temperature from cubicle to cubicle and from room to room—the various cubicles within the building are in a state of thermal disequilibrium. When we take new temperature readings several hours later, what shall we find? There will be very little variation in temperature among the cubicles within each single room, but there may still be large temperature variations *among* rooms. When we take readings again several days later, we find an almost uniform temperature throughout the building; the

[15] This discussion of near decomposability is based upon H. A. Simon and A. Ando, "Aggregation of Variables in Dynamic Systems," *Econometrica*, *29*: 111–138 (April 1961). The example is drawn from the same source, pp. 117–118. The theory has been further developed and applied to a variety of economic and political phenomena by Ando and F. M. Fisher. See F. M. Fisher, "On the Cost of Approximate Specification in Simultaneous Equation Estimation," *Econometrica*, *29*: 139–170 (April 1961), and F. M. Fisher and A. Ando, "Two Theorems on *Ceteris Paribus* in the Analysis of Dynamic Systems," *American Political Science Review*, *61*: 103–113 (March 1962).

temperature differences among rooms have virtually disappeared.

We can describe the process of equilibration formally by setting up the usual equations of heat flow. The equations can be represented by the matrix of their coefficients, r_{ij}, where r_{ij} is the rate at which heat flows from the ith cubicle to the jth cubicle per degree difference in their temperatures. If cubicles i and j do not have a common wall, r_{ij} will be zero. If cubicles i and j have a common wall and are in the same room, r_{ij} will be large. If cubicles i and j are separated by the wall of a room, r_{ij} will be nonzero but small. Hence, by grouping together all the cubicles that are in the same room, we can arrange the matrix of coefficients so that all its large elements lie inside a string of square submatrices along the main diagonal. All the elements outside these diagonal squares will be either zero or small (see Figure 7). We may take some small number, ϵ, as the upper bound of the extra-diagonal elements. We shall call a matrix having these properties a *nearly decomposable matrix*.

Now it has been proved that a dynamic system that can be described by a nearly decomposable matrix has the properties, stated earlier, of a nearly decomposable system. In our simple example of heat flow this means that in the short run each room will reach an equilibrium temperature (an average of the initial temperatures of its offices) nearly independently of the others; and that each room will remain approximately in a state of equilibrium over the longer period during which an over-all temperature equilibrium is being established throughout the building. After the intraroom short-run equilibria have been reached, a single thermometer in each room will be adequate to describe the dynamic behavior of the entire system—separate thermometers in each cubicle will be superfluous.

Near Decomposability of Social Systems

As a glance at Figure 7 shows, near decomposability is a rather strong property for a matrix to possess, and the matrices that have this property will describe very special dynamic systems—vanishingly few systems out of all

	A1	A2	A3	B1	B2	C1	C2	C3
A1	—	100	—	2	—	—	—	—
A2	100	—	100	1	1	—	—	—
A3	—	100	—	—	2	—	—	—
B1	2	1	—	—	100	2	1	—
B2	—	1	2	100	—	—	1	2
C1	—	—	—	2	—	—	100	—
C2	—	—	—	1	1	100	—	100
C3	—	—	—	—	2	—	100	—

Figure 7. A Hypothetical Nearly Decomposable System

In terms of the heat-exchange example of the text, A1, A2, and A3 may be interpreted as cubicles in one room, B1 and B2 as cubicles in a second room, and C1, C2, and C3 as cubicles in a third. The matrix entries then are the heat diffusion coefficients between cubicles.

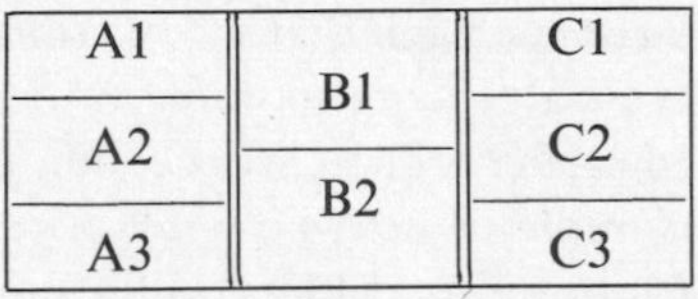

A1	B1	C1
A2		C2
A3	B2	C3

those that are thinkable. How few they will be depends, of course, on how good an approximation we insist upon. If we demand that epsilon be very small, correspondingly few dynamic systems will fit the definition. But we have already seen that in the natural world nearly decomposable systems are far from rare. On the contrary, systems in which each variable is linked with almost equal strength with almost all other parts of the system are far rarer and less typical.

In economic dynamics, the main variables are the prices and quantities of commodities. It is empirically true that the price of any given commodity and the rate at which it is exchanged depend to a significant extent only on the prices and quantities of a few other commodities, together with a few other aggregate magnitudes, like the average price level or some over-all measure of economic activity. The large linkage coefficients are associated, in general, with the main flows of raw materials and semi-finished products within and between industries. An input-output matrix of the economy, giving the magnitudes of these flows, reveals the nearly decomposable structure of the system—with one qualification. There is a consumption subsystem of the economy that is linked strongly to variables in most of the other subsystems. Hence, we have to modify our notions of decomposability slightly to accommodate the special role of the consumption subsystem in our analysis of the dynamic behavior of the economy.

In the dynamics of social systems, where members of a system communicate with and influence other members, near decomposability is generally very prominent. This is most obvious in formal organizations, where the formal authority relation connects each member of the organization with one immediate superior and with a small number of subordinates. Of course, many communications in organizations follow other channels than the lines of formal authority. But most of these channels lead from any particular individual to a very limited number of his superiors, subordinates, and associates. Hence, departmental boundaries play very much the same role as the walls in our heat example.

Physicochemical Systems

In the complex systems familar in biological chemistry, a similar structure is clearly visible. Take the atomic nuclei in such a system as the elementary parts of the system, and construct a matrix of bond strengths between elements. There will be matrix elements of quite different orders of magnitude. The largest will generally correspond to the covalent bonds, the next to the ionic bonds, the third group to hydrogen bonds, still smaller linkages to van der Waals forces.[16] If we select an epsilon just a

[16] For a survey of the several classes of molecular and intermolecular forces, and their dissociation energies, see Setlow and Pollard, *op. cit.*, Chapter 6. The energies of typical covalent bonds are of the order of 80–100 k cal/mole, of the hydrogen bonds, 10 k cal/mole. Ionic bonds generally lie between these two levels; the bonds due to van der Waals forces are lower in energy.

little smaller than the magnitude of a covalent bond, the system will decompose into subsystems—the constituent molecules. The smaller linkages will correspond to the intermolecular bonds.

It is well known that high-energy, high-frequency vibrations are associated with the smaller physical subsystems, low-frequency vibrations with the larger systems into which the subsystems are assembled. For example, the radiation frequencies associated with molecular vibrations are much lower than those associated with the vibrations of the planetary electrons of the atoms; the latter, in turn, are lower than those associated with nuclear processes.[17] Molecular systems are nearly decomposable systems, the short-run dynamics relating to the internal structures of the subsystems, the long-run dynamics to the interactions of these subsystems.

A number of the important approximations employed in physics depend for their validity on the near decomposability of the systems studied. The theory of the thermodynamics of irreversible processes, for example, requires the assumption of macroscopic disequilibrium but microscopic equilibrium,[18] exactly the situation described in our heat-exchange example. Similarly, computations in quantum mechanics are often handled by treating weak interactions as producing perturbations on a system of strong interactions.

Some Oberservations on Hierarchic Span

To understand why the span of hierarchies is sometimes very broad—as in crystals—and sometimes narrow, we need to examine more detail of the interactions. In general,

[17] Typical wave numbers for vibrations associated with various systems (the wave number is the reciprocal of wave length hence proportional to frequency):
steel wire under tension—10^{-10} to 10^{-9} cm^{-1}
molecular rotations—10^{0} to 10^{2} cm^{-1}
molecular vibrations—10^{2} to 10^{3} cm^{-1}
planetary electrons—10^{4} to 10^{5} cm^{-1}
nuclear rotations—10^{9} to 10^{10} cm^{-1}
nuclear surface vibrations—10^{11} to 10^{12} cm^{-1}.

[18] S. R. de Groot, *Thermodynamics of Irreversible Processes* (New York: Interscience Publishers, 1951), pp. 11–12.

the critical consideration is the extent to which interaction between two (or a few) subsystems excludes interaction of these subsystems with the others. Let us examine first some physical examples.

Consider a gas of identical molecules, each of which can form covalent bonds, in certain ways, with others. Let us suppose that we can associate with each atom a specific number of bonds that it is capable of maintaining simultaneously. (This number is obviously related to the number we usually call its valence.) Now suppose that two atoms join and that we can also associate with the combination a specific number of external bonds it is capable of maintaining. If this number is the same as the number associated with the individual atoms, the bonding process can go on indefinitely—the atoms can form crystals or polymers of indefinite extent. If the number of bonds of which the composite is capable is less than the number associated with each of the parts, then the process of agglomeration must come to a halt.

We need only mention some elementary examples. Ordinary gases show no tendency to agglomerate, because the multiple bonding of atoms "uses up" their capacity to interact. While each oxygen atom has a valence of two, the O_2 molecules have a zero valence. Contrariwise, indefinite chains of single-bonded carbon atoms can be built up, because a chain of any number of such atoms, each with two side groups, has a valence of exactly two.

Now what happens if we have a system of elements that possess both strong and weak interaction capacities and whose strong bonds are exhaustible through combination? Subsystems will form, until all the capacity for strong interaction is utilized in their construction. Then these subsystems will be linked by the weaker second-order bonds into larger systems. For example, a water molecule has essentially a valence of zero—all the potential covalent bonds are fully occupied by the interaction of hydrogen and oxygen molecules. But the geometry of the molecule creates an electric dipole that permits weak interaction between the water and salts dissolved in it—whence such phenomena as its electrolytic conductivity.[19]

[19] See, for example, L. Pauling, *General Chemistry* (San Francisco: W. H. Freeman, 2nd ed., 1953), Chapter 15.

Similarly, it has been observed that, although electrical forces are much stronger than gravitational forces, the latter are far more important than the former for systems on an astronomical scale. The explanation, of course, is that the electrical forces, being bipolar, are all "used up" in the linkages of the smaller subsystems, and that significant net balances of positive or negative charges are not generally found in regions of macroscopic size.

In social as in physical systems there are generally limits on the simultaneous interaction of large numbers of subsystems. In the social case, these limits are related to the fact that a human being is more nearly a serial than a parallel information-processing system. He can carry on only one conversation at a time, and although this does not limit the size of the audience to which a mass communication can be addressed, it does limit the number of people simultaneously involved in most other forms of social interaction. Apart from requirements of direct interactions, most roles impose tasks and responsibilites that are time consuming. One cannot, for example, enact the role of "friend" with large numbers of other people.

It is probably true that in social as in physical systems, the higher-frequency dynamics are associated with the subsystems, the lower-frequency dynamics with the larger systems. It is generally believed, for example, that the relevant planning horizon of executives is longer, the higher their location in the organizational hierarchy. It is probably also true that both the average duration of an interaction between executives and the average interval between interactions are greater at higher than at lower levels.

Summary: Near Decomposability

We have seen that hierarchies have the property of near decomposability. Intracomponent linkages are generally stronger than intercomponent linkages. This fact has the effect of separating the high-frequency dynamics of a hierarchy—involving the internal structure of the components—from the low-frequency dynamics—involving interaction among components. We shall turn next to some important consequences of this separation for the description and comprehension of complex systems.

The Description of Complexity

If you ask a person to draw a complex object—such as a human face—he will almost always proceed in a hierarchic fashion.[20] First he will outline the face. Then he will add or insert features: eyes, nose, mouth, ears, hair. If asked to elaborate, he will begin to develop details for each of the features—pupils, eyelids, lashes for the eyes, and so on—until he reaches the limits of his anatomical knowledge. His information about the object is arranged hierarchically in memory, like a topical outline.

When information is put in outline form, it is easy to include information about the relations among the major parts and information about the internal relations of parts in each of the suboutlines. Detailed information about the relations of subparts belonging to different parts has no place in the outline and is likely to be lost. The loss of such information and the preservation mainly of information about hierarchic order is a salient characteristic that distinguishes the drawings of a child or someone untrained in representation from the drawing of a trained artist. (I am speaking of an artist who is striving for representation.)

Near Decomposability and Comprehensibility

From our discussion of the dynamic properties of nearly decomposable systems, we have seen that comparatively little information is lost by representing them as hierarchies. Subparts belonging to different parts only interact in an aggregative fashion—the detail of their interaction can be ignored. In studying the interaction of two large molecules, generally we do not need to consider in detail the

[20] George A. Miller has collected protocols from subjects who were given the task of drawing faces and finds that they behave in the manner described here (private communication). See also E. H. Gombrich, *Art and Illusion* (New York: Pantheon Books, 1960), pp. 291–296.

interactions of nuclei of the atoms belonging to the one molecule with the nuclei of the atoms belonging to the other. In studying the interaction of two nations, we do not need to study in detail the interactions of each citizen of the first with each citizen of the second.

The fact, then, that many complex systems have a nearly decomposable, hierarchic structure is a major facilitating factor enabling to us understand, to describe, and even to "see" such systems and their parts. Or perhaps the proposition should be put the other way round. If there are important systems in the world that are complex without being hierarchic, they may to a considerable extent escape our observation and our understanding. Analysis of their behavior would involve such detailed knowledge and calculation of the interactions of their elementary parts that it would be beyond our capacities of memory or computation.[21]

I shall not try to settle which is chicken and which is egg: whether we are able to understand the world because it is hierarchic or whether it appears hierarchic because those aspects of it which are not elude our understanding and observation. I have already given some reasons for supposing that the former is at least half the truth—that evolving complexity would tend to be hierarchic—but it may not be the whole truth.

[21] I believe the fallacy in the central thesis of W. M. Elsasser's *The Physical Foundation of Biology*, mentioned earlier, lies in his ignoring the simplification in description of complex systems that derives from their hierarchic structure. Thus (p. 155):

"If we now apply similar arguments to the coupling of enzymatic reactions with the substratum of protein molecules, we see that over a sufficient period of time, the information corresponding to the structural details of these molecules will be communicated to the dynamics of the cell, to higher levels of organization as it were, and may influence such dynamics. While this reasoning is only qualitative, it lends credence to the assumption that in the living organism, unlike the inorganic crystal, the effects of microscopic structure cannot be simply averaged out; as time goes on this influence will pervade the behavior of the cell 'at all levels.'"

But from our discussion of near decomposability, it would appear that those aspects of microstructure that control the slow developmental aspects of organismic dynamics can be separated out from the aspects that control the more rapid cellular metabolic processes. For this reason we should not despair of unraveling the web of causes. See also J. R. Platt's review of Elsasser's book in *Perspectives in Biology and Medicine*, *2*: 243–245 (1959).

Simple Descriptions of Complex Systems

One might suppose that the description of a complex system would itself be a complex structure of symbols—and indeed, it may be just that. But there is no conservation law that requires that the description be as cumbersome as the object described. A trivial example will show how a system can be described economically. Suppose the system is a two-dimensional array like this:

$$\begin{matrix} A & B & M & N & R & S & H & I \\ C & D & O & P & T & U & J & K \\ M & N & A & B & H & I & R & S \\ O & P & C & D & J & K & T & U \\ R & S & H & I & A & B & M & N \\ T & U & J & K & C & D & O & P \\ H & I & R & S & M & N & A & B \\ J & K & T & U & O & P & C & D \end{matrix}$$

Let us call the array $\begin{vmatrix} AB \\ CD \end{vmatrix}$ a, the array $\begin{vmatrix} MN \\ OP \end{vmatrix}$ m, the array $\begin{vmatrix} RS \\ TU \end{vmatrix}$ r, and the array $\begin{vmatrix} HI \\ JK \end{vmatrix}$ h. Let us call the array $\begin{vmatrix} am \\ ma \end{vmatrix}$ w, and the array $\begin{vmatrix} rh \\ hr \end{vmatrix}$ x. Then the entire array is simply $\begin{vmatrix} wx \\ xw \end{vmatrix}$. While the original structure consisted of 64 symbols, it requires only 35 to write down its description:

$$S = \begin{matrix} wx \\ xw \end{matrix}$$

$$w = \begin{matrix} am \\ ma \end{matrix} \qquad x = \begin{matrix} rh \\ hr \end{matrix}$$

$$a = \begin{matrix} AB \\ CD \end{matrix} \qquad m = \begin{matrix} MN \\ OP \end{matrix} \qquad r = \begin{matrix} RS \\ TU \end{matrix} \qquad h = \begin{matrix} HI \\ JK \end{matrix}$$

We achieve the abbreviation by making use of the redundancy in the original structure. Since the pattern $\begin{matrix} AB \\ CD \end{matrix}$, for example, occurs four times in the total pattern, it is economical to represent it by the single symbol, *a*.

If a complex structure is completely unredundant—if no aspect of its structure can be inferred from any other—then it is its own simplest description. We can exhibit it, but we cannot describe it by a simpler structure. The hierarchic structures we have been discussing have a high degree of redundancy, hence can often be described in economical terms. The redundancy takes a number of forms, of which I shall mention three:

1. Hierarchic systems are usually composed of only a few different kinds of subsystems, in various combinations and arrangements. A familiar example is the proteins, their multitudinous variety arising from arrangements of only twenty different amino acids. Similarly, the ninety-odd elements provide all the kinds of building blocks needed for an infinite variety of molecules. Hence, we can construct our description from a restricted alphabet of elementary terms corresponding to the basic set of elementary subsystems from which the complex system is generated.

2. Hierarchic systems are, as we have seen, often nearly decomposable. Hence only aggregative properties of their parts enter into the description of the interactions of those parts. A generalization of the notion of near decomposability might be called the "empty world hypothesis" —most things are only weakly connected with most other things; for a tolerable description of reality only a tiny fraction of all possible interactions needs to be taken into account. By adopting a descriptive language that allows the absence of something to go unmentioned, a nearly empty world can be described quite concisely. Mother Hubbard did not have to check off the list of possible contents to say that her cupboard was bare.

3. By appropriate "recoding," the redundancy that is present but unobvious in the structure of a complex system can often be made patent. The commonest recoding of descriptions of dynamic systems consists in replacing a description of the time path with a description of a differential law that generates that path. The simplicity, that is, resides in a constant relation between the state of the system at any given time and the state of the system

a short time later. Thus, the structure of the sequence 1 3 5 7 9 11 . . . is most simply expressed by observing that each member is obtained by adding 2 to the previous one. But this is the sequence that Galileo found to describe the velocity at the end of successive time intervals of a ball rolling down an inclined plane.

It is a familar proposition that the task of science is to make use of the world's redundancy to describe that world simply. I shall not pursue the general methodological point here, but I shall instead take a closer look at two main types of description that seem to be available to us in seeking an understanding of complex systems. I shall call these *state description* and *process description*, respectively.

State Descriptions and Process Descriptions

"A circle is the locus of all points equidistant from a given point." "To construct a circle, rotate a compass with one arm fixed until the other arm has returned to its starting point." It is implicit in Euclid that if you carry out the process specified in the second sentence, you will produce an object that satisfies the definition of the first. The first sentence is a state description of a circle, the second a process description.

These two modes of apprehending structures are the warp and weft of our experience. Pictures, blueprints, most diagrams, and chemical structural formulas are state descriptions. Recipes, differential equations, and equations for chemical reactions are process descriptions. The former characterize the world as sensed; they provide the criteria for identifying objects, often by modeling the objects themselves. The latter characterize the world as acted upon; they provide the means for producing or generating objects having the desired characteristics.

The distinction between the world as sensed and the world as acted upon defines the basic condition for the survival of adaptive organisms. The organism must develop correlations between goals in the sensed world and actions in the world of process. When they are made conscious and verbalized, these correlations correspond to what we usually call means-end analysis. Given a desired state of affairs and an existing state of affairs, the task of an adaptive organism is to find the difference between these two states and then to find the correlating process that will erase the difference.[22]

Thus, problem solving requires continual translation between the state and process descriptions of the same complex reality. Plato, in the *Meno*, argued that all learning is remembering. He could not otherwise explain how we can discover or recognize the answer to a problem unless we already know the answer.[23] Our dual relation to the world is the source and solution of the paradox. We pose a problem by giving the state description of the solution. The task is to discover a sequence of processes that will produce the goal state from an initial state. Translation from the process description to the state description enables us to recognize when we have succeeded. The solution is genuinely new to us—and we do not need Plato's theory of remembering to explain how we recognize it.

There is now a growing body of evidence that the activity called human problem solving is basically a form of means-end analysis that aims at discovering a process description of the path that leads to a desired goal. The general paradigm is: given a blueprint, to find the corresponding recipe. Much of the activity of science is an application of that paradigm: given the description of some natural phenomena, to find the differential equations for processes that will produce the phenomena.

The Description of Complexity in Self-Reproducing Systems

The problem of finding relatively simple descriptions for complex systems is of interest not only for an understanding of human knowledge of the world but also for an

[22] See H. A. Simon and A. Newell, "Simulation of Human Thinking," in M. Greenberger (ed.), *Management and the Computer of the Future* (New York: Wiley, 1962), pp. 95–114, esp. pp. 110 ff.

[23] *The Works of Plato*, B. Jowett, translator (New York: Dial Press, 1936), Vol. 3, pp. 26–35.

explanation of how a complex system can reproduce itself. In my discussion of the evolution of complex systems, I touched only briefly on the role of self-reproduction.

Atoms of high atomic weight and complex inorganic molecules are witnesses to the fact that the evolution of complexity does not imply self-reproduction. If evolution of complexity from simplicity is sufficiently probable, it will occur repeatedly; the statistical equilibrium of the system will find a large fraction of the elementary particles participating in complex systems.

If, however, the existence of a particular complex form increased the probability of the creation of another form just like it, the equilibrium between complexes and components could be greatly altered in favor of the former. If we have a description of an object that is sufficiently clear and complete, we can reproduce the object from the description. Whatever the exact mechanism of reproduction, the description provides us with the necessary information.

Now we have seen that the descriptions of complex systems can take many forms. In particular, we can have state descriptions, or we can have process descriptions—blueprints or recipes. Reproductive processes could be built around either of these sources of information. Perhaps the simplest possibility is for the complex system to serve as a description of itself—a template on which a copy can be formed. One of the most plausible current theories, for example, of the reproduction of deoxyribonucleic acid (DNA) proposes that a DNA molecule, in the form of a double helix of matching parts (each essentially a "negative" of the other), unwinds to allow each half of the helix to serve as a template on which a new matching half can form.

On the other hand, our current knowledge of how DNA controls the metabolism of the organism suggests that reproduction by template is only one of the processes involved. According to the prevailing theory, DNA serves as a template both for itself and for the related substance ribonucleic acid (RNA). RNA, in turn, serves as a template for protein. But proteins—according to current knowledge—guide the organism's metabolism not by the template method but by serving as catalysts to govern reaction rates in the cell. While RNA is a blueprint for protein, protein is a recipe for metabolism.[24]

Ontogeny Recapitulates Phylogeny

The DNA in the chromosomes of an organism contains some, and perhaps most, of the information that is needed to determine its development and activity. We have seen that, if current theories are even approximately correct, the information is recorded not as a state description of the organism but as a series of "instructions" for the construction and maintenance of the organism from nutrient materials. I have already used the metaphor of a recipe; I could equally well compare it with a computer program, which is also a sequence of instructions, governing the construction of symbolic structures. Let me spin out some of the consequences of the latter comparison.

If genetic material is a program—viewed in its relation to the organism—it is a program with special and peculiar properties. First, it is a self-reproducing program; we have already considered its possible copying mechanism. Second, it is a program that has developed by Darwinian evolution. On the basis of our watchmaker's argument, we may assert that many of its ancestors were also viable programs—programs for the subassemblies.

Are there any other conjectures we can make about the structure of this program? There is a well-known generalization in biology that is verbally so neat that we would be reluctant to give it up even if the facts did not support it: ontogeny recapitulates phylogeny. The individual organism, in its development, goes through stages that resemble some of its ancestral forms. The fact that the human embryo develops gill bars and then modifies them for other purposes is a familiar particular belonging to the generalization. Biologists today like to emphasize the qualifications of the principle—that ontogeny recapitulates only the grossest aspects of phylogeny, and these only crudely.

[24] C. B. Anfinsen, *The Molecular Basis of Evolution* (New York: Wiley, 1959), Chapters 3 and 10, will qualify this sketchy, oversimplified account. For an imaginative discussion of some mechanisms of process description that could govern molecular structure, see H. H. Pattee, "On the Origin of Macromolecular Sequences." *Biophysical Journal, 1*: 683–710 (1961).

These qualifications should not make us lose sight of the fact that the generalization does hold in rough approximation—it does summarize a very significant set of facts about the organism's development. How can we interpret these facts?

One way to solve a complex problem is to reduce it to a problem previously solved—to show what steps lead from the earlier solution to a solution of the new problem. If, around the turn of the century, we wanted to instruct a workman to make an automobile, perhaps the simplest way would have been to tell him how to modify a wagon by removing the singletree and adding a motor and transmission. Similarly, a genetic program could be altered in the course of evolution by adding new processes that would modify a simpler form into a more complex one—to construct a gastrula, take a blastula and alter it!

The genetic description of a single cell may, therefore, take a quite different form from the genetic description that assembles cells into a multicelled organism. Multiplication by cell division would require, as a minimum, a state description (the DNA, say), and a simple "interpretive process"—to use the term from computer language—that copies this description as a part of the larger copying process of cell division. But such a mechanism clearly would not suffice for the differentiation of cells in development. It appears more natural to conceptualize that mechanism as based on a process description, and a somewhat more complex interpretive process that produces the adult organism in a sequence of stages, each new stage in development representing the effect of an operator upon the previous one.

It is harder to conceptualize the interrelation of these two descriptions. Interrelated they must be, for enough has been learned of gene-enzyme mechanisms to show that these play a major role in development as in cell metabolism. The single clue we obtain from our earlier discussion is that the description may itself be hierarchical, or nearly decomposable, in structure, the lower levels governing the fast, "high-frequency" dynamics of the individual cell, the higher-level interactions governing the slow, "low-frequency" dynamics of the developing multicellular organism.

There are only bits of evidence, apart from the facts of recapitulation, that the genetic program is organized in this way, but such evidence as exists is compatible with this notion.[25] To the extent that we can differentiate the genetic information that governs cell metabolism from the genetic information that governs the development of differentiated cells in the multicellular organization, we simplify enormously—as we have already seen—our task of theoretical description. But I have perhaps pressed this speculation far enough.

The generalization that, in evolving systems whose descriptions are stored in a process language, we might expect ontogeny partially to recapitulate phylogeny has applications outside the realm of biology. It can be applied as readily, for example, to the transmission of knowledge in the educational process. In most subjects, particularly in the rapidly advancing sciences, the progress from elementary to advanced courses is to a considerable extent a progress through the conceptual history of the science itself. Fortunately, the recapitulation is seldom literal—any more than it is in the biological case. We do not teach the phlogiston theory in chemistry in order later to correct it. (I am not sure I could not cite examples in other subjects where we do exactly that.) But curriculum revisions that rid us of the accumulations of the past are infrequent and painful. Nor are they always desirable—

[25] There is considerable evidence that successive genes along a chromosome often determine enzymes controlling successive stages of protein syntheses. For a review of some of this evidence, see P. E. Hartman, "Transduction: A Comparative Review," in W. D. McElroy and B. Glass (eds.), *The Chemical Basis of Heredity* (Baltimore: Johns Hopkins Press, 1957), at pp. 442–454. Evidence for differential activity of genes in different tissues and at different stages of development is discussed by J. G. Gall, "Chromosomal Differentiation," in W. D. McElroy and B. Glass (eds.), *The Chemical Basis of Development* (Baltimore: Johns Hopkins Press, 1958), at pp. 103–135. Finally, a model very like that proposed here has been independently, and far more fully, outlined by J. R. Platt, "A 'Book Model' of Genetic Information Transfer in Cells and Tissues," in M. Kasha and B. Pullman (eds.), *Horizons in Biochemistry* (New York: Academic Press, forthcoming). Of course, this kind of mechanism is not the only one in which development could be controlled by a process description. Induction, in the form envisaged in Spemann's organizer theory, is based on process description, in which metabolites in already formed tissue control the next stages of development.

partial recapitulation may, in many instances, provide the most expeditious route to advanced knowledge.

Summary: The Description of Complexity

How complex or simple a structure is depends critically upon the way in which we describe it. Most of the complex structures found in the world are enormously redundant, and we can use this redundancy to simplify their description. But to use it, to achieve the simplification, we must find the right representation.

The notion of substituting a process description for a state description of nature has played a central role in the development of modern science. Dynamic laws, expressed in the form of systems of differential or difference equations, have in a large number of cases provided the clue for the simple description of the complex. In the preceding paragraphs I have tried to show that this characteristic of scientific inquiry is not accidental or superficial. The correlation between state description and process description is basic to the functioning of any adaptive organism, to its capacity for acting purposefully upon its environment. Our present-day understanding of genetic mechanisms suggests that even in describing itself the multicellular organism finds a process description—a genetically encoded program—to be the parsimonious and useful representation.

Conclusion

Our speculations have carried us over a rather alarming array of topics, but that is the price we must pay if we wish to seek properties common to many sorts of complex systems. My thesis has been that one path to the construction of a nontrivial theory of complex systems is by way of a theory of hierarchy. Empirically, a large proportion of the complex systems we observe in nature exhibit hierarchic structure. On theoretical grounds we could expect complex systems to be hierarchies in a world in which complexity had to evolve from simplicity. In their dynamics, hierarchies have a property, near decomposability, that greatly simplifies their behavior. Near decomposability also simplifies the description of a complex system and makes it easier to understand how the information needed for the development or reproduction of the system can be stored in reasonable compass.

In both science and engineering, the study of "systems" is an increasingly popular activity. Its popularity is more a response to a pressing need for synthesizing and analyzing complexity than it is to any large development of a body of knowledge and technique for dealing with complexity. If this popularity is to be more than a fad, necessity will have to mother invention and provide substance to go with the name. The explorations reviewed here represent one particular direction of search for such substance.

INDEX

A

B

C

D

E

F

G

H

O

P

Q

R

S